Securing Digital Video

Eric Diehl

Securing Digital Video

Techniques for DRM and Content Protection

Springer

Eric Diehl
Technicolor, Security & Content
Protection Labs
Ave. Belle Fontaine 1
Rennes 35576
France

ISBN 978-3-642-17344-8 ISBN 978-3-642-17345-5 (eBook)
DOI 10.1007/978-3-642-17345-5
Springer Heidelberg New York Dordrecht London

Library of Congress Control Number: 2012940230

ACM Computing Classification: E.3, H.5.1, K.6.5, K.4.4

Printed on acid-free paper

Springer is part of Springer Science+Business Media (www.springer.com)

Acknowledgments

First of all, I would like to thank my wife for her constant support for this work and for her invaluable patience.

I would like to thank many colleagues and friends who carefully reviewed portions of the manuscript; their comments made this book more readable and attractive. First, my colleagues at Technicolor, including Michael Arnold, Jean-Marc Boucqueau, Gwenael Doerr, Alain Durand, Marc Eluard, Marc Joye, Andrew Hackett, Sylvain Lelievre, Mohamed Karroumi, Yves Maetz, Antoine Monsifrot, and Charles Salmon-Legagneur. Then my gratitude goes to friends, including Jukka Alve (SESKO), Olivier Bomsel (Ecole des Mines), Ton Kalker (DTS), Kevin Morgan (Arxan), Florian Pestoni (Adobe), Nicolas Prigent (Supelec), Arnaud Robert (Disney), Rei Safavi-Naini (University of Calgary), and Massimo Savazzi (Microsoft).

I would like to offer special thanks to Prof. Jean-Jacques Quisquater and Prof. Benoit Macq of the Université Catholique de Louvain. They made me believe that I could write such a book. Most probably, without them I would not have dared to begin this work.

Many thanks to Technicolor who allowed me to use illustrations from its bank of images.

I am grateful to Ronan Nugent and Springer, who showed keen interest in this book.

Eric Diehl

Contents

Chapter 1
Introduction

Content protection and Digital Rights Management (DRM) are among those fields that get a lot of attention from content owners, consumers, scholars, and journalists. Content owners require content protection and DRM to protect and maximize their revenues. Consumers fear that they may have to always purchase a new version of their preferred movies for each new generation of consumer electronics device (video tape, DVD, mp4, and Blu-ray). They also fear that content owners will spy on their viewing habits. Scholars are afraid that DRM may be a barrier to knowledge sharing. Journalists follow the consumers' trends and sometimes describe these techniques as devilish or crap (Content Restriction, Annulment, and Protection [1]). Content protection and DRM have a bad reputation

Many solutions are already deployed in the market, adding to the confusion of consumers and even journalists. Figure 1.1 displays a non-exhaustive list of technologies protecting audiovisual contents. Unfortunately, many of these solutions are not interoperable. Consumers are confronted with restricting technologies that do not bring to them any perceived benefits.

Furthermore, with content protection and DRM being related to security, corresponding available information is sometimes scarce. Restricting information is often a survival precaution for security. Unfortunately, in our world of freely flowing information, lack of public information is often perceived as serving evil purposes [2].

There was a need for a book that would describe the foundations of all these systems and that would detail some of the most emblematic instantiations of content protection and DRM. The objective of this book is to demystify the technologies used by these video content protections so that the reader can rationally make up his mind.

Several sources of information are available: scientific publications, white papers, specifications, information disclosed under non-disclosure agreements, articles written by news reporters, public forums, due diligence meetings, actual use of the solutions, and reverse-engineering. Although this book only discloses publicly available information, I employed all these sources. Non-public information

E. Diehl, *Securing Digital Video*, DOI: 10.1007/978-3-642-17345-5_1,

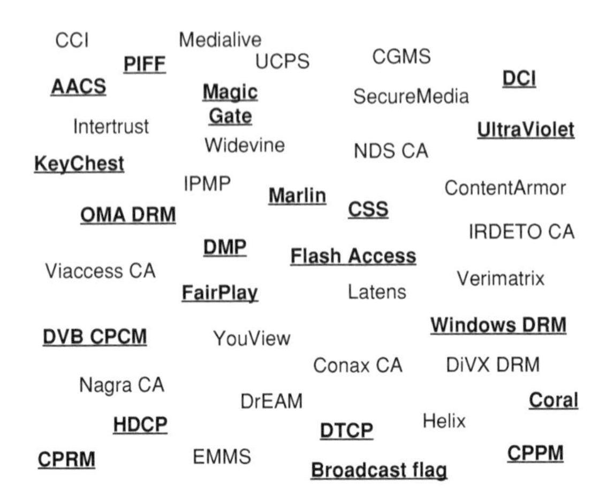

Fig. 1.1 The galaxy of content protection (This book describes in detail the technologies in underlined bold characters. UCPS is the Chinese equivalent of HDCP. CONAX, IRDETO, NAGRA, NDS, and VIACCESS are conditional access system providers. IPMP is a framework for DRM in MPEG4 and MPEG21, which has not been deployed. ContentArmor is a professional DRM for B2B developed by Technicolor. EMMS, Helix, Intertrust, and Latens are DRM technology providers. YouView is the content protection framework of the BBC)

provides us with deep knowledge of the actual product. This deeper insight helps better describe these systems, even when using solely public information. There is a controversial source of public information: hacker sites. Hackers have sometimes reverse-engineered proprietary systems and have published their findings. This book provides pointers to these works, especially if the hacked version is obsolete or if the information is widely available.

The first part of the book builds the foundations. Chapter 2 provides the rationales for protecting digital video, and gives a quick tour of video piracy. Chapter 3 briefly describes the toolbox used. This chapter is a short introduction to cryptography, watermarking, tamper-resistance, and rights expression languages. The reader familiar with these techniques may skip this chapter. Chapter 4 shows different ways to model video content protection and DRM. It introduces the four-layer model that will be used in the remaining chapters to describe the behavior of the systems. Using the same model will highlight the strong similarities among the systems.

The second part of the book describes the main deployed solutions. Chapter 5 introduces the video ecosystem. Chapter 6 explains how the broadcast world protects video. Chapter 7 introduces so-called DRM, describing systems such as Microsoft DRM, and Apple's FairPlay. Chapter 8 elucidates how pre-recorded content such as that on DVD or Blu-ray is protected. Chapter 9 details the future methods used to protect content within the home network. Chapter 10 describes the protection scheme of digital cinema. Systems whose specifications are publicly available are described with more details than are proprietary ones.

The third part of the book looks at the future and at exploratory topics. Chapter 11 explores the hot problem of DRM interoperability. Chapter 12 explores some challenges and fascinating research topics.

The Devil's in the Details
Security is complex. A minor implementation mistake may ruin a theoretically secure solution. Inserts entitled "The Devil's in the Details" illustrate the gap between theory and real-world security. We describe some hacks to highlight what went wrong. Each insert concludes with a lesson that security practitioners have learnt (sometimes the hard way). It should give a glimpse of the work of designers of content protection solutions.

This book focuses on protection of video content. DRM technology protects many other types of content such as eBooks, documents, and games. This book does not describe solutions used to protect these types of content. Nevertheless, the foundations of these DRMs are identical to those used by video content protection. Thus, readers should easily be able to extrapolate how these other types of content are protected.

Chapter 2
Why Protect Video?

2.1 Introduction

Digital video is a set of binary information that represents video content in compressed or uncompressed format. Bits can be easily copied, transferred, or stored. This is one of the main advantages of digital video over analog video. Nevertheless, digital video is intrinsically unprotected, i.e., digital video can be easily viewed by anybody and be infinitely duplicated. Thus, in some cases, it needs protection, or it needs to control access under some defined conditions.

Providing a universal definition of video content protection and DRM is probably as difficult as embarking on the quest for the Holy Grail. The actual scope of video content protection is so broad, encompasses so many aspects, and affects so many stakeholders that no single definition may satisfy everybody. An efficient video content protection system has the right equilibrium of three main aspects: technology, business model, and legal constraints. Video content protection is like a three-legged stool [3].[1] For many reasons, this equilibrium is extremely difficult to find (Fig. 2.1):

- The expectations of consumers and content providers do not necessarily align, and sometimes even diverge [4]. Thus, consumers sometimes reject new business models. Nevertheless, content providers recognize that consumers expect privacy, anonymity, and transferability [5] and try to find business models acceptable to consumers.
- Copyright laws are complex and not universal. Even in national laws, some concepts are fuzzy, such as in the US fair use concept. Dematerialized content creates even more confusion, at least for consumers. Which copyright law applies to content created in the US, stored on Slovenian servers, and accessed by a French user?

---

[1] The metaphor of the stool has its limits. Indeed, a three-legged stool is rather stable as long as the difference in the lengths of the three legs is not too much. The equilibrium is more difficult in case of content protection. Nevertheless, if one leg is missing, the entire system collapses.

E. Diehl, *Securing Digital Video*, DOI: 10.1007/978-3-642-17345-5_2,

Fig. 2.1 The equilibrium of DRM

- Technology has limitations and cannot perfectly map the copyright laws or perfectly fulfill business models.

Furthermore, each of these three components interacts together. Consumers will only endorse DRM if it is acceptable to them [6]. Content owners will offer digital content only if they are convinced that legislation will permit them to pursue infringers.

We will propose a set of complementary definitions and issues that should encompass all aspects of video content protection. A digital video may represent a work protected by some copyright laws. Thus, some technical means should enforce these laws (Sect. 2.2). The digital video may have commercial value, which should also be preserved by technical means (Sect. 2.3). In the context of digital video, piracy is the act either of infringing copyright laws or of violating commercial relationships.

The following sections examine in detail the copyright issues and the business issues and give a brief overview of piracy.

2.2 Copyright Issues

Copyright laws give to a creator the monopoly of exclusive control of any form of performance and reproduction of his work. Performance means the action of publicly exhibiting any form of the opus, whereas reproduction means duplicating the opus on a physical or electronic medium. Copyright or intellectual property laws define the rules under which contributive works must be rewarded [7]. The aims of copyright laws are multiple. First, they should protect the interests of the author. Above all, it should be an incentive for creation. Being able to hope for commercial gain is essential to an author, thereby justifying investment in content creation.[2] If only fame or recognition interests the author, copyrights under Creative Commons provide also an incentive to publish whilst granting a minimal level of protection. Second, it should protect the interests of the investors. Producers and editors will not provide any financial and technical support to the

[2] The same goal is behind the concept of a patent. The inventor may expect to get some royalties if third parties use his invention. This retribution is an incentive to share knowledge for the common good.

author if they cannot expect a return on investment [8]. Third, public interests have to be also preserved. It is fundamental that consumers have fair access to cultural goods. The initial objective of copyright was to incentivize the creation of cultural goods and also to better educate people. This is why the rights granted by copyright are valid for a fixed period of time. Once this period finished, the work enters the public domain and can be freely used. This balance between private and public interests is at the heart of the difficult tradeoff of copyright laws.

Often, the first copyright law is attributed to the English Queen Anne, with the Statute of Queen Anne on 10 April 1710. It awarded the author the right to control copies for a period of 14 years, renewable for another 14-year period. In 1777, the French author Pierre Augustin Caron de Beaumarchais founded the Société des Auteurs et Compositeurs Dramatiques (SACD). The French SACD still collects copyright levies for its members [9]. In 1790, the United States enacted its first copyright law with a 14-year protection for *American* authors only. The objective was only to protect US intellectual property [10]. American publishers could shamelessly and freely copy foreign authors. In 1886, the Bern convention promulgated a first international treaty signed by many nations such as France, Germany, the UK, and Spain but not the US. In 1952, the symbol © was created. Its use is widely adopted worldwide. In 1996, the World Intellectual Property Organization (WIPO) established a new international treaty, the WIPO Copyright Treaty [11], which defined obligations concerning technological measures (Article 11) and rights management of information (Article 12).

There are no "universal" copyright laws. Although the WIPO Copyright Treaty is an international treaty that many countries ratified,[3] it is only a loose framework defining some common principles that are not accurate. For instance, Article 12 defines the obligations concerning rights management of information:

> (1) Contracting Parties shall provide adequate and effective legal remedies against any person knowingly performing any of the following acts knowing, or with respect to civil remedies having reasonable grounds to know, that it will induce, enable, facilitate or conceal an infringement of any right covered by this Treaty or the Berne Convention:
>
> (i) to remove or alter any digital rights management information without authority;
> (ii) to distribute, import for distribution, broadcast or communicate to the public, without authority, works or copies of works knowing that digital rights management information has been removed or altered without authority.

Despite the existence of the WIPO treaty, each country has its own regulations, and implements/interprets the treaty in its own way. For instance, since October 28, 1998, the US had the Digital Millennium Copyright Act (DMCA) [7]. In 2001, the European Union issued the European Union Copyright Directive (EUCD) [13]. Nevertheless, every European country has its own transcription of this directive in its national legislation. Therefore, due to these multiple definitions, video protection should ideally be able to adapt to each country's local requirements. This

[3] At the time of writing, 88 countries have enforced this treaty [12].

heterogeneity of copyright laws makes it difficult to prosecute international infringers in other countries. In 2003, the author of DeCSS, a piece of software that could crack the protection of DVDs (Sect. 8.2, CSS), was acquitted by Norwegian courts [14]. Most probably, this would not have been the case if he had been a US citizen. DMCA does not hold in Norway and the Motion Picture Association of America (MPAA) did not find an equivalent in Norwegian legislation.

Human laws are never Boolean. Copyright laws are the same way. They introduce many exemptions such as personal use, news reporting, quotation and criticism, educational use, archival storage, and library privilege. For instance, DMCA introduces the concept of fair use. Fair use is the right to make, in some circumstances, copies, or modifications of a work without asking for permission from the author of the original work, provided the original author is cited. US fair use defines a list of exceptions such as criticism, comments, news reporting, teaching, scholarship, and research. DMCA vaguely states these exceptions and they evolve along time. Every three years, the Librarian of Congress considers new exemptions. In July 2010, the Librarian of Congress added six new exemptions to DMCA [15]. The six new exemptions were:

- extracting short video sequences from protected DVDs for creating works for non-commercial purposes or criticism,
- modifying protected mobile phone applications to ensure interoperability with other handsets,
- jail breaking mobile phones to enable their operation on other operator networks,
- hacking video games for evaluating or enhancing their security,
- hacking software protected by a dongle if it is malfunctioning or obsolete,
- hacking eBooks if no available edition offers text to voice feature.

It is currently understood that to determine if fair use is applicable, the following factors should be checked [16]:

1. the purpose and the character of the usage, including whether such use is of commercial nature or for nonprofit educational purposes, i.e., use of commercial nature is not compatible with fair use,
2. the nature of the copyrighted work,
3. the amount and substantiality of the portion used in relation to the copyrighted work as whole, as there is a difference between using 1 and 90 % of a work, and DMCA does not define such a threshold, leaving the judge to define it,
4. the effect of the use upon the potential market of the original work and on its value.

The interpretations of these exemptions result in countless fierce litigations in US courts. Lawyers have difficulties in interpreting the law; how then can we expect that computer programmers, or, worse, computer programs, could successfully accomplish such fuzzy interpretation [17]? It is unlikely that any computer-based system could accommodate the shadow aspects of fair use [18].

To illustrate this problem, let us analyze a classical dispute: the rights of the consumer to make a backup of his/her copy. This is allowed by the fair use doctrine of DMCA and by the French “copie privée” (private copy) [19]. How can a computer differentiate between a legitimate private copy and an illegal copy? The difference is not technical. The difference is in the intent of the user. This is the classical dilemma when using anti-ripping techniques for pre-recorded media (Sect. 8.1, Anti-ripping). In July 1985, a French law, called "loi Lang" [20], attempted to solve the issue by applying a copy levy to every recordable media. This tax encompasses optical recordable media such as CD-R, CD-RW, DVD-R, and DVD-RW, and static storage such as flash memories and embedded hard drives.[4] For instance, every mp3 player is subject to this tax. This tax is, for instance, about 7 euros (i.e., about $9) for an iPhone. The collected levy is redistributed to copyright owners, distributors, and authors to compensate for illegal copies [21]. The levy is only redistributed to audio and video copyright owners with a share proportional to their commercial market share.[5] Although this system has the advantage of being simple, it has many critics. Some people feel that the levy is unfair because it imposes the tax on every use of recordable media, even legitimate ones that do not involve copyright content. For instance, if you backup your computer onto a DVD-R, you will have to pay the tax regardless whether your backup contains copyrighted content or not. Furthermore, some people complain that this tax gives a white hand to pirates and, worse, that it legitimizes illegal downloads. In fact, the levy takes into account the estimated level of piracy. A recent survey claimed that 40 % of the content stored on recordable media was coming from Peer-to-Peer (P2P) sharing networks.[6] If the levy takes into account piracy then it covers both a private copy (of a legally acquired content) and an illegal copy (of P2P downloaded content). Thus, according to the opponents, a P2P download should not be anymore considered illegal because its effect is already integrated in the tax. Furthermore, users may conclude that paying a levy on everything is equivalent to acquiring an unlimited license to copy, in other words it may be an incentive to engage in piracy [22]. Already, many people wrongly believe that the high cost of subscriptions to ISPs is a toll for downloading. Many people admit that they embraced high-speed Internet to download content [23].

Another example is the lending/renting of a piece of content. In the US, it falls into the realm of the "doctrine of first sale". This doctrine allows the purchaser to do whatever she desires with her purchase. Thus, she may resell it or lend it. This is often not possible with DRM-protected content. Numerous judicial actions and laws have introduced exceptions. No rationale seems to drive these exceptions. For instance, in the US, you may rent audio tapes, but not music CDs. You may photocopy articles of a newspaper that you purchased, but if you make

---

[4] This levy was already applied on analog audio and video tapes.

[5] In other words, it is assumed that the more an author is successful, the more this work will be pirated.

[6] As a simplification, I will often refer to P2P as an illegal network. I would like to underscore that P2P technology per se is not illegal. P2P technology has many legal and beneficial usages. Unfortunately, the largest current usage of P2P networks is illegal sharing of copyrighted content. The same argument holds for direct download (DDL) technology.

photocopies of articles of a newspaper purchased by your company for its library that is illegal [24].

Thus, many opponents of DRM claim that DRM technologies do not protect copyright but rather extend the scope of the copyright to the realm of where, when, and how a piece of content can be used. This extension of scope was not the original purpose of copyright laws [17].

Video protection should attempt to enforce these rewarding mechanisms. Here is one of the first difficulties of video protection. As copyright laws have two aspects, their definition and their interpretation, it is impossible for content protection to implement a smart, adaptive interpretation of the law. It can only implement a limited, rigid interpretation.

2.3 Business Issues

Digital goods have two characteristics: non-rivalry and non-excludability [25]. A good is non-rivalrous if its consumption by one person does not limit its consumption by other persons. For instance, the book you are reading now is non-rivalrous. Once you have finished it (and I hope appreciated it), it will still be available to you or anybody else. This is not true of the soda you drank while reading the book. Once drunk, your soda is unavailable to anybody else. A good is non-excludable when it is impossible to prevent someone from consuming it. Ambient air is a non-excludable good. There is no way to restrict its usage (except, of course, by strangulation). Water is an excludable good. It is possible to block it with a dam. There are two forms of non-excludability: the collective use of a good and the copying of a good. For instance, all the members of a dormitory can watch a movie in the community room (the first type of non-excludability). Every member of this dormitory may duplicate the same movie (the second type of non-excludability). Rivalry and excludability are not linked.

Digital goods are in essence non-excludable and non-rivalrous. When Alice reads a digital file from a server she does not prevent Bob from reading the same file (non-rivalry). If Alice, or rather her software, knows the semantics and syntax of the digital file, then she can read the file (non-excludability). She may even make a bit-to-bit copy of the file and redistribute it. Video protection techniques are technological tools that enforce excludability of information goods, which otherwise would be public goods. Video protection techniques artificially generate scarcity of digital content, thus increasing its commercial value.

In the analog age, the cost of the physical media and the cost of replication inherently created scarcity. For instance, consumers could duplicate their vinyl discs only by audio recording. High-quality audio tapes were more expensive than regular ones. Furthermore, the copied audio track was not as good as the original one. Successive copies were worse. With the advent of CD and CD-R, the cost of the recordable support decreased and ease of duplication, with the proliferation of personal computers, increased. Scarcity was diminishing. With the advent of audio and video compression, P2P sharing, Direct Download sites, and portable players, scarcity vanished.

Business plans for copyrighted works are based on sequential but separate releases, i.e., managed and controlled scarcity. Versioning in economics is the form of discrimination in which customers are induced by pricing to select different versions of a product [26]. Book publishing is a typical example of versioning. The hardcover version of a book is more expensive than the paperback version, although both versions have exactly the same text.

Digital goods, being non-excludable, implicitly are not ready for versioning. Video protection techniques can become a versioning tool. Through conditional access, it can provide for each content a predefined set of utilities: the right to view (hear), modify, record, excerpt, translate into another language, keep for a certain period, distribute, etc. It is then possible to sell at different prices the same digital piece of content with different utilities. For instance, the view-once version of a movie should be cheaper than the same movie on a DVD. Versioning content increases the revenues generated from one original work and offers a price more adjusted to a consumer's willingness to pay.

A good illustration of versioning is the release window system deployed by the US movie industry. Figure 2.2 illustrates the distribution of different versions of the same movie. The different windows are as follows:

- *Theatrical Release*: The public life of a movie starts with its theatrical release. This is when a movie starts to be projected in theaters. The theatrical release date may not be the same for every country. The main reason for this discrimination is economic. Each country requires a version in its own language. Replacing the original audio tracks with foreign language tracks is called dubbing. In some cases, subtitles are added. These operations are costly and lengthy. Starting the dubbing and subtitling processes, called internationalization, after domestic theatrical release reduces the initial investment. The incomes coming from theaters may help the industry to recoup the cost of internationalization.

  Furthermore, dubbing requires a full version of the movie. Thus, it is a perfect target for pirates.[7] A full version of a movie before theatrical release is the most valuable material for pirates. The earlier the copy leaks, the greater its value on the pirate market. Thus, starting internationalization after the first theatrical release reduces the risk of early leakage.

  English speakers living outside the US may be eager to get access as early as possible to blockbusters. In the absence of immediate legal availability in their country, they may turn to illegal distribution channels such as P2P channels. Thus, studios have adapted their strategy. If a title has a high probability of becoming a huge worldwide commercial success, the studio will program a worldwide release.

[7] The version used by professional dubbing teams is often of low resolution to limit the risk of leakage. Nevertheless, this version is still better than most early camcorded pirated versions.

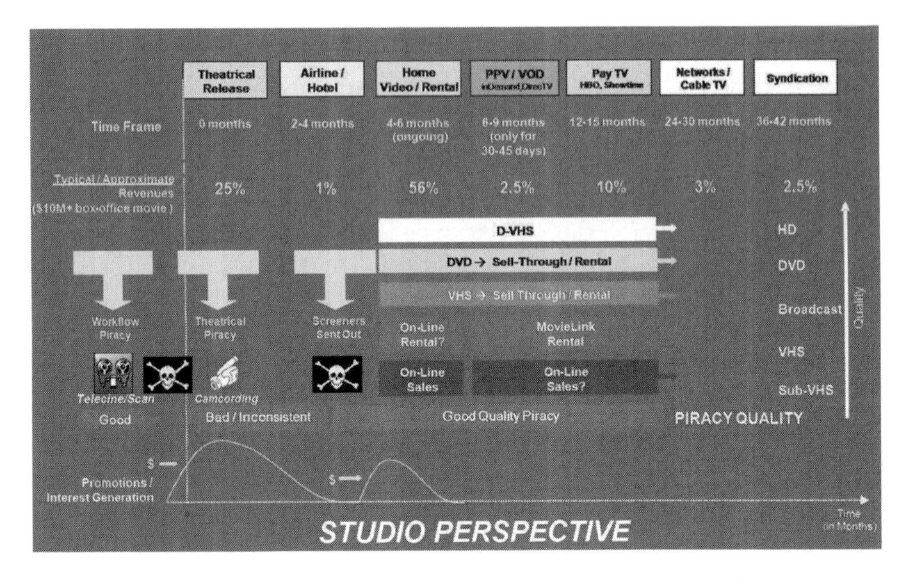

Fig. 2.2 Release windows (courtesy Technicolor)

The movie will be launched at the same date in theaters all around the world. This strategy mitigates at least the second type of piracy.

- *Hospitality Release:* One month after theatrical release, a version of the movie may be distributed on airplanes and hotel pay channels.
- *Home/Video Release:* Depending on its success, 4–6 months later, the movie will be available for purchase as a DVD or for rental. The less successful a movie is, the earlier its DVD release will be. Currently, the market trend is to try to shrink this delay. This release window is currently the most profitable one. In 2009, packaged video represented worldwide revenues of about $43 billion while theatrical revenues were about $30 billion [27]. Therefore, preserving this window is paramount for studios.
- *Video on Demand (VOD) Release*: Two months later, the movie appears for sale on pay per view (PPV) or VOD. The customer can purchase the movie either online through the Internet or through the PPV offer of broadcasters such as HBO PPV and Showtime.
- *Pay TV Release*: About one year after its theatrical release, Pay TV channels will first propose the movie to their subscribers.
- *Premium Channel*: Typically, dedicated network channels or premium channels offer the movie two years later.
- *Syndication Release*: Later, the movie will be available for Free to Air TV. A title remains in the catalog of a broadcaster for many years. In fact, broadcasters purchase the rights to play it several times. This is why they regularly program old blockbusters. It secures their returns on investments.

Only legal and technical enforcements entitle this distribution strategy. There is some form of loose coupling between the two types of enforcement. Contractors sign legal agreements. They define constraints for both parties. The owner grants some rights for using a piece of content under given conditions. Sometimes, these constraints explicitly mandate some technologies to enforce the agreed contractual rules. For instance, if a US content provider signs an exclusivity contract for distributing its TV shows to a French broadcaster, then the US content provider has the obligation to turn down any French viewer visiting its free TV catch-up site. This is why viewers not based in the US may get a screen displaying the following message when attempting to access some pieces of content: "THE VIDEO YOU ARE TRYING TO WATCH CANNOT BE VIEWED FROM YOUR LOCATION OR COUNTRY".

Versioning may in fact benefit consumers. Versioning enabled by DRM allows a lower price. If a content provider can sell the same content with different commercial packages, it may propose a more elaborate offering to customers. This is called price discrimination. Let us take a basic example. The content provider wants to sell a movie as one-time viewing or unlimited viewing. If it proposes one-time viewing for price a and unlimited viewing for price b, a consumer may assume that $a \ll b$. People interested in watching the movie only once will be willing to pay less than people wanting to watch it multiple times. Were the content provider not able to propose (and thus enforce) one-time viewing, consumers would only have access to the unlimited offer, i.e., the one with the higher price tag b. This would not be acceptable to people wanting to watch it only once. The impact of versioning on the price is already visible on some current commercial offers: EMI sold, through iTunes, DRM-protected songs for $0.99 and the same DRM-free song for $1.29 [28]. This segmentation of the commercial offer is only possible due to DRM. Without DRM, consumers would only have one offer: the most expensive one. Thus, DRM can provide a means for offering price differentiation in a bearable way for consumers [29]. Price discrimination benefits both consumers and content owners. Although they are most probably not aware of it, consumers are used to price discrimination of physical goods. Hardcover books come out earlier than paperback reprints and are more expensive. Buying a DVD is more expensive than renting it. Even in software, the full feature application is more expensive than the "lite" one [30].

The ultimate versioning strategy would be to propose to Alice an individualized price according to her profile. Michael Lesk, from Rutgers University, attempted to answer why online music stores sell to every customer one product at the same price [29]. Interestingly, on the one hand, every song is sold at the same price, regardless of the fame of its performer or the success of the song on the billboard chart. On the other hand, the price of the corresponding CD varies according to the artist's fame. Online stores, which sell individual tracks, should have a rather good knowledge of Alice's profile. Thus, they could easily propose a personalized price slightly lower than the maximum price she would be ready to pay. According to Lesk, it is not the consumer's privacy but her remorse that frightens the sellers. Merchants believe that Alice would be upset if she learnt that Bob purchased the same song at a lower price in the same digital store. Nevertheless, DRM could offer such price discrimination.

Similarly, often when a national broadcaster purchases the commercial rights for a title, they are only granted for a given geographical footprint. When TV broadcasts were only terrestrial, this was not an issue. Of course, people living near borders had access to foreign channels, but this was marginal. With the advent of satellite broadcast, this became an issue. A satellite broadcast has a much wider coverage than a terrestrial one. Satellite broadcast wiped out the frontiers. Satellite broadcasters have the choice to buy the right either for the potential audience in the full area covered by the satellite, or for the audience in their own country. In the latter case, satellite broadcasters have to implement DRM to restrict access to the program only to their customers. This is one of the reasons behind the birth of modern Conditional Access Systems (CAS).

Note that copyright does not necessarily imply a business relationship. The Creative Commons is an example of where copyright is asserted without any monetary retribution [31].

2.4 A Brief Overview of Digital Piracy

In every battle, it is wise to know the enemy [32]. Thus, we will study different aspects of piracy. First, we describe the pirates and their organization. In a second part, we provide some data about the estimated cost of video piracy.

It is useful to understand the illegal distribution chain, sometimes called the Scene or the Darknet [33]. There are mainly two types of piracy, depending on the delivered goods: physical goods (such as DVDs) or digital goods.

Piracy of physical goods represents the market of illegally pressed or burnt discs. Often referred to as bootleg CDs or DVDs, they are sold in shops, on the streets, and even on websites. The quality of these goods is irregular, varying from extremely good discs to poor ones (poor video quality and bad media). This is a major form of piracy; in particular, in Russian or Asian markets. This piracy is mainly driven by organized crime and terrorist organizations [34]. For instance, Indian "D" Company, the terrorist organization allegedly controlled by Ibrahim Dawood,[8] fully controls all aspects of piracy of Bollywood movies [35]. Several facts explain this new interest by organized crime in the piracy of digital goods. The financial and technical entry barriers are extremely low. Enforcement risk is low. The sentence for DVD counterfeiting is a few years whereas the sentence for drug dealing exceeds ten years. Furthermore, the benefit margins generated by DVD smuggling are far higher than those from drug dealing. In other words, bootleg piracy is less dangerous and more lucrative than other forms of crime [36] and offers the best return on investment (ROI).

Piracy of digital goods corresponds to illegal electronic exchange of content through sharing networks such as P2P networks and Direct Download (DDL) sites

---

[8] Ibrahim Dawood is the alleged organizer of the 1993 Mumbai bombings. He is the most wanted criminal in India.

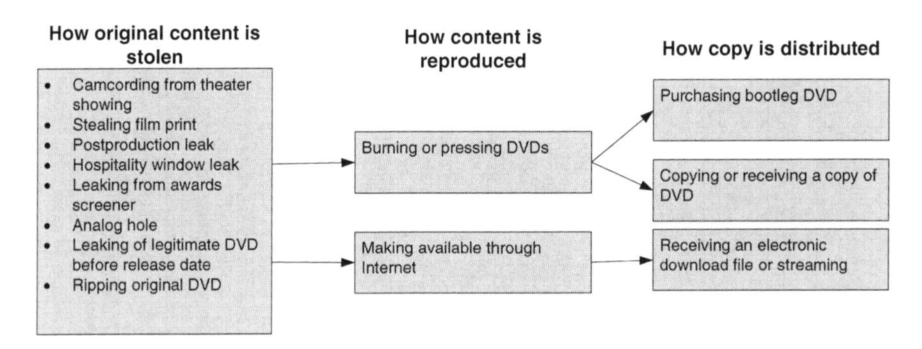

Fig. 2.3 Stages and forms of film piracy

that propose streaming content (such as RapidShare and Megaupload). Here also, the quality of the video varies from early camcorder capture, with bad quality picture and people walking in front of the screen, to bit-to-bit pristine copy of a DVD or a Blu-ray disc. This form of piracy is extremely *en vogue* in the Western market. It gets also the largest media coverage.

Figure 2.3 describes how pirates steal, reproduce, and distribute illegal content. The first step is the theft of the content itself. There are many potential sources of leakage. The most worrying one is camcording in theaters, with varying quality. Another source is ripping the original DVD either before the release date or once the content is in the official public market. The quality is excellent. Nevertheless, there are many other sources such as film theft (intercepting it during transfer to the theater). According to a study, more than 60 % of the early content found on the Darknet was due to insiders [37].

Let us now study how a small number of people can flood the world with illegal copies of legitimate content. Figure 2.4 illustrates the pyramid of piracy for P2P [38]. This pyramid can be extrapolated to other distribution chains such as illegal DVDs. There are four levels of actors:

- *The suppliers* are the people who actually capture or steal the content. For early release, i.e., before release of normal DVD, it may be the team of people who go to the theater and capture the video. Many samples display people walking in front of the camera. The capture conditions are poor. Sometimes, capture is done from the projection booth with the complicity of an operator after regular display hours. In that case, the quality of the capture may be excellent. Sometimes audio tracks are recorded from seats equipped for hearing-impaired people, thus offering better quality than that from live capture in the theater.
  Of course, other suppliers exist, such as people receiving early DVDs for reviewing or for academy awards, known as screeners. This is a worrisome source of early leakage [39]. Sometimes, an insider may steal a movie in postproduction, or while it is being subtitled.
- *The release teams* are the people who actually package the pirated good. They may perform many processing operations on the supplied video. We will

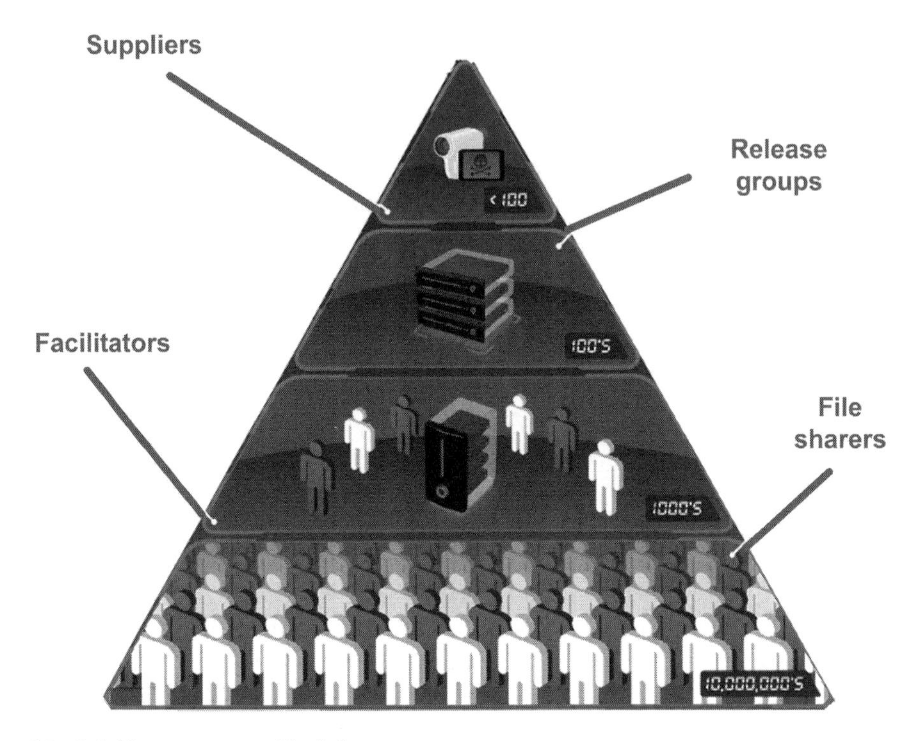

Fig. 2.4 The scene: pyramid of piracy

describe later in this chapter the organization of such teams. Some release teams come from the cyber culture phenomena of *warez*. Warez appeared in the 1980s with the advent of Bulletin Board Systems.[9] Warez are groups of crackers who collect every piece of "protected" data available on the Internet [40]. Initially, warez members collected and cracked software and offered it to other members. Cracking the protection of software was one of the challenges for fame among the crackers. To be widely recognized by the community, one had to be the first to offer a cracked version of a piece of software. Offering a cracked version the same day as its official release (0-day warez) was reckoned to be a great exploit. Offering a cracked version before its official release date (negative-day warez) was *THE* exploit. The ultimate goal and source of pride of warez was to have the largest available collection of software, including expensive professional applications such as AutoCAD and PhotoShop. Later, warez extended their scope to multimedia data such as movies and songs. In the warez culture, specific names categorize the collected data. This taxonomy is widely reused by sharing sites. The following categories describe stolen multimedia content.

[9] Bulletin Board Systems (BBS) were the first sharing systems. A BBS is mainly a computer connected to one or several phone lines. It was possible to remotely contact this server, load and download files. The Internet has mostly phased out BBS (except in some Third World countries).

- *Movies* are pirated movies that have been either camcorded in theaters or ripped from DVDs. They are mainly available in compressed format such as DiVX and XviD. They are packaged in containers such as `avi`, `mkv` (Matroska), `mov` (QuickTime), and `mp4` (MPEG4). More rarely, they are available as a pristine bit-to-bit copy of the DVD. In that case, they are packaged as an `iso` file.
- *MP*3s hold the pirated albums and songs. The preferred compression format is mp3, but other formats such as FLAC (FlawLess Audio Compression) with extension flc are available, although more scarcely.
- *MVids* are music videos captured from TV or ripped from DVDs and Blu-ray discs.
- *Subs* contain files for subtitles. They have to be used with files from Movies and TV-rips.
- *TV-rips* hold shows captured from TV. The response time of hackers for this category is extremely short. It is not rare to find an episode of a TV show one hour after it was broadcast and a few hours later with the corresponding subtitles.

• The *facilitators* are the people who contribute to the distribution. There are mainly two types of distribution channels: physical goods, such as DVDs and digital files. It is interesting to observe the path of dissemination of digital files. Often the early releases are only available on dedicated ftp sites. Access to these sites is limited to very few co-opted people. To enter these controlled circles, the candidate has to provide new content. These releases often have high commercial value and end up in physical replication premises. It is rather common to find illegal DVDs on the streets two or three days after theatrical release. Later these releases appear on dedicated Internet Relay Chat (IRC) forums. Here also, access to the IRC is controlled and limited to known trusted members. Eventually, the release appears on P2P networks or streaming sites. Nevertheless, currently the dissemination path is accelerating and it is rather easy to find early releases on P2P networks [41].[10] Recent studies have proven that most pieces of content available on P2P sites are provided by a limited number of facilitators [42]. They often hide behind public proxies, anonymizing networks such as Tor, and even Virtual Private Networks (VPN). Specialized P2P networks such as OMEMO offer some form of anonymity [43]. The new trend is to use "Stock & Share" or DDL sites such as RapidShare and Megaupload. These sites store files that are uploaded by users. The files can be accessed via a regular http link accessible with a regular browser. Such links are found using searching engines like Google, or exchanged via tweets. Usually, the free service gives a limited download bandwidth for one unique download. The paid service provides unlimited bandwidth for an unlimited number of simultaneous downloads. Some DDL sites offer monetary incentives to share valuable content.

---

[10] This acceleration may be partly explained by the sharing networks moving from free model to business models, for instance, based on advertisements.

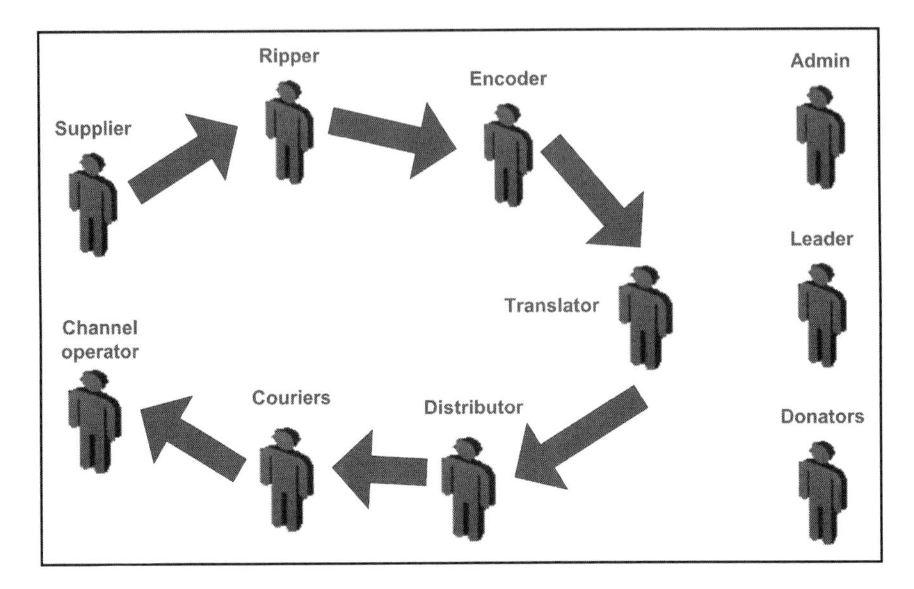

Fig. 2.5 Release team

The more people who download the content you uploaded, the more advantages for you.[11] As for P2P, DDL sites have legitimate usage and are useful to exchange large files. Unfortunately, most of the files stored by users on these sites are illegally reproduced copyrighted files.

In the case of physical goods, the facilitators are the replicators and the sellers. In large-scale piracy, they use organized distribution channels similar to commercial ones.

- The *users/file sharers* are members of the crowd who "benefit" from previous teams. Most of them will use P2P clients such as emule, vuze, and μtorrent to download, upload, and share files or directly view content from streaming sites whose addresses they found through leech sites.

  As illustrated by Fig. 2.5, the organization of release teams is extremely structured.

[11] For instance, Megaupload rewarded its uploading members up to $10,000 if their file was downloaded more than 5,000,000 times [44]. It should be noted that their terms of service banned illegal content. They stated: "You agree to not use Megaupload's Service to: **c.** upload, post, e-mail, transmit or otherwise make available any Content that infringes any patent, trademark, trade secret, copyright or other proprietary rights of any party". However, when shutting down the Megaupload site on 19th January 2012, the FBI claimed among the many indictments that "The conspirators further allegedly offered a rewards program that would provide users with financial incentives to upload popular content and drive Web traffic to the site, often through user-generated websites known as linking sites. The conspirators allegedly paid users whom they specifically knew uploaded infringing content and publicized their links to users throughout the world".

- The *supplier* provides the original material, for instance, screeners. Those screeners are copies of a movie sent before actual public release. Recipients of such screeners are newspaper critics or members of the Oscar academy. Obviously, those screeners have excellent video quality. They are potential sources of early leakages [45, 46].
  In some cases, the supplier may only deliver audio tracks. Often, the audio capture is done from seats equipped for hearing impaired persons. Thus, the quality of the audio tracks is excellent.
- The *ripper* possibly removes the copy protection of the content provided by the supplier. Then he reformats the content. He may resize the image to avoid black bars, or slightly remove keystone effects.[12]
- The *encoder* compresses the content in a smaller file. DivX is currently the preferred compression package for illegal distribution.
- The *translator* adds subtitles. In some cases, there is even dubbing, i.e., adding multiple audio tracks with different languages. It is common to find US series in English with foreign subtitles. In some cases, the translator appends to already captured video the foreign audio tracks captured in a theater [47]. Video is language-independent whereas an audio track is language-dependent.
  Therefore, you may sometimes find a movie with credits in one language and audio in another language. In fact, it is less dangerous and easier to record a sound track than it is to capture the actual movies. Thus, sometimes release groups use this lower risk strategy. A release team may dub content provided by another release team.
- The *distributor* performs the final packaging, for instance, through replication. It may sometimes be on an industrial scale.
- *Couriers* bring the illegal package to the channel operators, previously called facilitators. Operators will post them on ftp servers, seed them on P2P networks, and upload them to cyberlockers (DDL sites). Then they will publicize them on dedicated IRCs for the ftp version, on tracker sites for P2P networks, on twitter using tools such as Torrent Tweet [48], and on leech sites for cyberlockers. The use of ftp servers is not for a large audience and serves mainly commercial transactions such as providing a copy for bootlegs. Access to the ftp site is closely controlled.
- Release teams often have a leader and an administrator who supervise the technical work.
- Donators may provide resources by donating money, either directly by purchasing content or indirectly by providing material.

Piracy is not only due to organized teams. Some form of piracy is accessible to non-technical people, for instance, DVD ripping. When done on a small scale, for instance, for neighbors and friends, it is called ant piracy [49]. The company Rovi,

[12] A keystone effect is the trapezoidal distortion that occurs when the capturing device is not strictly parallel to the screen. This is often the case with wild capture in theaters.

previously known as Macrovision, proposes an interesting classification of the people who rip DVDs [50]. It proposes a four-scale ladder:

- Level 1 rippers use basic tools to copy and remove CSS. If they fail to rip a title, they will not try to design a new circumventing tool. They represent 60 % of rippers.
- Level 2 rippers use the same tools as level 1 rippers. If they fail to rip a title, they may investigate a solution through Google and may even post questions on dedicated forums. They represent 25 % of rippers.
- Level 3 rippers use combinations of ripping tools. If they fail to rip a title, they will know where to find solutions on forums. They do not post questions on forums and do not share their knowledge. They represent 10 % of rippers.
- Committed rippers are very technical. They answer on online forums. They may be beta testers of new versions of ripping tools. They represent 5 % of rippers.

We will illustrate the efficiency of piracy with "Star Trek" as an example. "Star Trek" was simultaneously available in theaters in selected countries on 6 May 2009. Two days later, the movie was officially released worldwide. The first bootleg DVD hit the street on 8 May 2009. It was a reasonably fair quality camcorder capture coming from a Russian theater. Release teams quickly internationally dubbed this Russian version. Five other camcorder captures appeared, by order of appearance on the Internet, from the Philippines, Ukraine, Spain, Germany, and the US. The Ukrainian copy was of excellent quality and soon became the reference material. At the end of August, more than five million IP addresses downloaded it from P2P sites. There is no estimate of the number of downloads coming from DDL streaming sites. At the same time, 35 million viewers watched "Star Trek" in theaters [51].

Obviously, digital piracy has a real impact on business. Unfortunately, despite numerous studies, a recent publication highlighted that it was impossible to estimate the real financial loss [52]. Undoubtedly, digital piracy is reaching an impressive level of professionalism. Some sites that sell illegal content are better looking and offer more interesting features than legitimate selling sites [53]. Furthermore, pirated content may have some perceived advantages for consumers that may not be present in legal content. For instance, some people complain about the unskippable trailers present in commercial DVDs that are absent in bootlegs [54]. Such arguments may give a false justification for pirating. Digital piracy started with a free business model in the early days of warez and P2P, went through an advertisement-based business model period with the proliferation of major popular torrent sites such as The Pirate Bay and Mininova,[13] and switched to a paid service model with DDL and VPN enabled P2P such as The Pirate Bay's iPredator. The goal is clearly to move towards a more business-oriented vision.

---

[13] Both sites are top 120 websites.

Chapter 3
A Tool Box

3.1 Different Needs, Different Tools

As highlighted by the previous chapter, digital video content protection has many aspects, and therefore uses many techniques. Ideally, it should incorporate four types of protection [55], which are complementary and serve the unique purpose of mitigating the misuse of a piece of content. Together they ensure strong protection. Their goals are the following (Fig. 3.1):

- control access to the asset,
- protect the asset itself,
- use forensics to trace the asset, and
- scout to limit the illegal use of the asset.

3.1.1 Controlling Access

The first barrier involves controlling physical or logical access to the asset. This barrier is obvious for professional activities (such as Business-to-Business (B2B), postproduction facilities and broadcast studios). Only authorized users should be authorized near the asset. This protective measure may take the form of physical control such as guards at the entrance or gates controlled by badges, biometric identification, and vaults. Video cameras may also monitor entrances and critical areas of the site.

In the digital world, the second type of access control is IT security. Typically, the IT department defines a perimeter that it defends against intruders by using firewalls, demilitarized zones, and virtual private networks. Within this secure perimeter, IT controls access to data according to access control lists and role-based policies. Access control-based protection is historically the first protection of a document. It derived from the seminal Bell-Lapadula security model [56], which defined hierarchical access control, and was refined over time. This type of

E. Diehl, *Securing Digital Video*, DOI: 10.1007/978-3-642-17345-5_3,

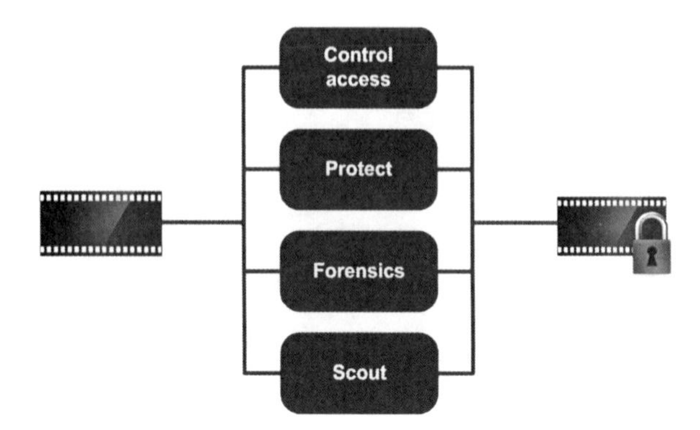

Fig. 3.1 Four goals of security

protection is outside the scope of this book but the interested reader may check "Access Control Systems" [57], "Practical UNIX and Internet Security" [58], and "Hacking exposed" [59] for further information.

In the Business-to-Customer (B2C) domain, some methods will prevent easy access to the asset. Consumer electronics devices may use System-on-Chips (SoC) (Sect. 3.5.2, Different Types of Components) or hide connections under epoxy layers.

3.1.2 Protecting the Asset

The second barrier targets direct attacks on the asset such as theft, alteration, and replacement. The associated tools are based on encryption and cryptographic signatures (Sect. 3.2, Cryptography). Encryption enforces confidentiality of the asset whereas signature ensures the integrity of data but also authenticates principals. Key management defines the ways to distribute and protect the secrets used by cryptography. If those secrets leak, then encrypted and signed assets are vulnerable. Most video content protection systems vary in their key management.

Cryptography is a solution to turn digital content, which is normally non-rival, into a rival good.

3.1.3 Forensic Marking

The third barrier complements the previous one. Eventually, any digital content has to be decrypted in order to be rendered in the analog world. At this point, the protection provided by encryption collapses. Forensic marking is used to insert, within the rendered content, information about the rendering source. The dedicated technology to perform this is watermarking (Sect. 3.3, Digital Watermark).

Forensic technologies can turn digital content, which is normally non-excludable, into an excludable good. It makes it possible for content owners to sue people who organized the illegal distribution.

3.1.4 Thwarting Illegal Distribution

Unfortunately, content can always leak. As a result, a fourth barrier is needed to attempt to limit the losses that are incurred.

The first step is to detect illegal content. The most efficient technology to do so is fingerprinting. A reference database stores a collection of fingerprints that contain unique characteristics about the content [60]. These fingerprints may be based on characteristics such as visual hashes, color information, time characteristics and points of interests. When the system spots a suspect piece of content, it extracts the relevant fingerprints and compares them to those stored in the reference database. Fingerprinting is superior to identification using cryptographic hash values because it is robust against geometrical modifications, mash-ups, and camcording.

Once an illegal piece of content is identified, corrective actions are taken depending on the context. Currently, some User Generated Content (UGC) websites filter out submitted material when they identify it as being copyrighted [61]. In the case of P2P and DDL platforms, a take-down notification may be sent to the sharers.

As the last resort, illegal dissemination has to be slowed down. The first objective is to deter the downloaders by providing the wrong piece of content instead of the expected one. Typical techniques spread decoys, fake content, or even encrypted content that requires a payment [62]. The second objective is to inhibit access either by slowing down the bandwidth or by routing the requester to controlled peers. No studio officially announced employing these kinds of techniques.

Obviously, this fourth goal encompasses also police action, such as raiding bootleg providers and legal litigation. This book does not describe in detail how to thwart illegal distribution. That would require a book in itself.

3.2 Cryptography

3.2.1 Introduction

Cryptography is arguably the most important technology for video content protection. This section introduces the basic concepts of cryptography so that the reader understands the remainder of the book. The reader who is already familiar with cryptography may skip this section and jump directly to Sect. 3.3. Although modern cryptography involves a lot of mathematics, this section will not dwell on the mathematical details. The reader interested in the mathematics can easily find relevant textbooks [63, 64].

Before walking through cryptography concepts, I would like to introduce four famous characters who will follow us in this book. They are often encountered in security-related publications.

- Alice and Bob are very talkative persons. They always exchange messages. They attempt to have secure transfer, i.e., that their messages cannot be understood by anybody other than themselves and nobody can modify these messages.
- Eve wishes to learn what Alice and Bob exchange. Thus, she is always eavesdropping on them. She is a passive attacker.
- Charlie also wants to learn what Alice and Bob exchange. He is an active attacker compared to Eve.

Let us now jump to cryptography. Cryptography serves several purposes:

- *Data Confidentiality*: Where the aim is to prevent Eve from understanding a message exchanged between Alice and Bob. For that, they will use encryption. Ideally, if Eve does not have the right secret decryption key, even if she knows the algorithm used, she should not be able to learn a single bit of the clear message from the encrypted message.

In fact, stronger security assumes a more constraining scenario for the defenders. An example is the chosen ciphertext attack. The hypothesis is that Charlie can select the ciphertext and obtains the decryption, while not knowing the secret key. The goal of Charlie is to obtain the key.

- *Authenticity:* Where the aim is to prove the identity of a principal. Bob has to be sure that he speaks to Alice. Mathematically, the principal proves that it knows or owns a secret. Based on the number of "secrets" involved in the protocol, we speak of multi-factor authentication. For instance, ideally to authenticate Alice, we should use many factors of authentication, implying [65]:
  - Something she knows, typically a secret such as a password or a passphrase.
  - Something she owns, for instance, a physical key, a token, or a smart card.
  - Something she is; usually, it uses biometrics such as voice recognition, fingerprint recognition, or retina recognition.
  - Some of her skills, for instance, her ability to recognize people, translate words into foreign language, or solve problems. (This last category is still in the experimental stage. Captcha is a good illustration of this type of authentication [66, 67]. Captchas are automatic challenges that only a human being should be able to solve, such as badly written sequences of characters. Here, the objective is not to authenticate an individual but to authenticate the fact that a principal is a human being rather than a machine or a piece of software. Captchas assess skill by assuming that a machine cannot successfully solve the challenge. Graphical passwords are another example of skill-based authentication [68].)

Digital content protection mainly uses one-factor authentication, based on the knowledge of one secret.

- *Integrity:* Where the aim is to prove that Charlie did not modify the message sent by Alice before Bob received it. Enforcement of integrity uses tools such as cryptographic hash functions.

There are two types of cryptosystems: symmetric (also called secret key) cryptography and asymmetric (also called public key) cryptography. This book uses the notational convention that $f_{\{k\}}(x)$ means the result of function f applied to value x using key K.

3.2.2 Symmetric Cryptography

Symmetric cryptosystems were the first historically established cryptographic systems. They were already in use in antiquity [69]. Alice and Bob share a common secret key K. To securely send a message m to Bob, Alice applies to message m an algorithm E, called encryption, using the key K as a parameter. The result $c = E_{\{k\}}(m)$ is called the ciphertext whereas m is often referred as plaintext. To retrieve plaintext m from the ciphertext c, Bob applies an algorithm D, called decryption, using the same key K such that $D_{\{k\}}(c) = D_{\{k\}}\left(E_{\{k\}}(m)\right) = m$.

The encryption and decryption algorithms are designed such that, at least in theory, without the knowledge of the secret key K, it would be infeasible for Eve to find m from c. More precisely, Eve should not learn any information whatsoever about a plaintext m given its ciphertext c beyond the length of the plaintext. In the nineteenth century, Auguste Kerckhoffs, a Dutch cryptographer, established the following rule, which is still in use: "A [cryptographic system] must require secrecy, and should not create problems should it fall into the enemy's hands." In other words, the robustness of a cryptographic system should rely on the secrecy of the key and not on the secrecy of its algorithm [70]. Thus, the usual assumption is that Eve has access to all the details of the encryption algorithm E, of the decryption algorithm D, and of the ciphertext c (Fig. 3.2).

The design of such an encryption/decryption algorithm is a delicate task. Its robustness relies on the complexity of the transformations applied to the data. It is a complex combination of transformations such as permutations, substitutions, and binary operations.

Eve always has a straightforward attack strategy at hand: the brute-force attack. In a brute-force attack, Eve explores every possible value for the key K. Thus, if K is k bits long, Eve will have to try at most 2^k values. If the size of the key is too small, then Eve can explore all the values in reasonable time. Today, experts estimate that the minimal key size of a symmetric cryptosystem is about 100 bits [71, 72]. Current practice is to use a key of at least 128 bits, which means at most 2^{128} trials (more than 3×10^{38} trials!). Even with the computing power of all the

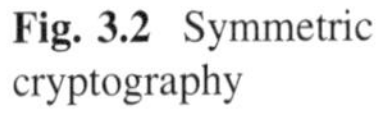
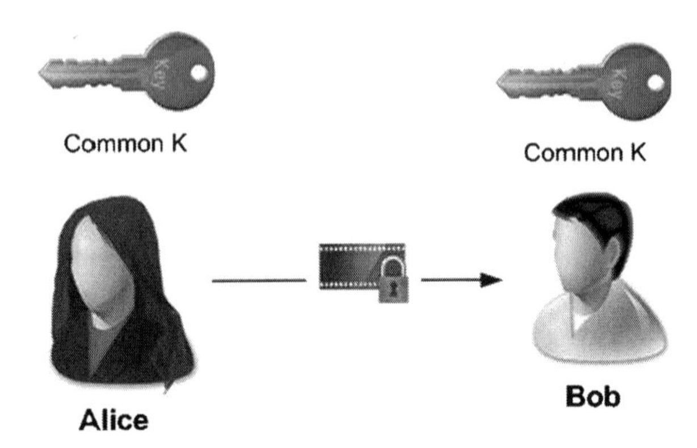

Fig. 3.2 Symmetric cryptography

currently available computers in the world, Eve would need several tens of years to explore these 2^{128} trials.

The first widely used and standardized symmetric cryptosystem was the Data Encryption Standard (DES). DES is a symmetric block cipher, i.e., the clear text is a block of bytes. IBM designed DES in the 1970s. The US National Institute of Standards and Technology (NIST) adopted DES in January 1977. It is a 64-bit block cipher using a 56-bit key.[1] Due to continuously increasing computing power, DES weakened. In 1998, a $250,000 dedicated machine broke by brute-force attacks a 56-bit key in an average time of 2.5 days [73]. Triple DES was DES's successor. It extended the actual size of the key to 112 bits, placing it beyond the reach of brute-force attacks.

To replace the aging DES, in January 1997, NIST and the US National Security Agency (NSA) launched a public contest for a new symmetric encryption algorithm. More than thirty candidates answered the challenge. For four years, the cryptographic community thoroughly cryptanalyzed these candidate algorithms ending up with a final selection of five algorithms. In 2000, from the five remaining algorithms, NIST selected Rijndael, better known now as the Advanced Encryption Standard (AES). AES is a block cipher with a key of 128, 192, or 256 bits [74]. It has been specifically designed to be fast both in hardware and software implementations, whereas the old DES was poorly performing in pure software implementation. 128-bit AES is currently the most widely used symmetric cryptosystem. AES is the first cryptographic algorithm that was standardized and selected using an open public challenge. Other examples of symmetric cryptography include Blowfish, TwoFish, and RC5.

Symmetric cryptography can also authenticate and guarantee the integrity of a message. For that purpose, it uses a Message Authentication Code (MAC) and a shared key K. Alice calculates for her message m the signature $\sigma = MAC_{\{k\}}(m)$.

---

[1] Initially, the US National Security Agency required decreasing the key size from 112 bits to its current size in order to decrease its strength. Many national regulations treat cryptographic algorithms as war weapons. Later in this book, we will see how the limitations imposed by such regulations have dramatically impacted some deployed video content protection systems.

She sends both m and σ. Bob verifies the authenticity and integrity of the received message m by verifying that the received σ satisfies the previous equation. To be secure, the MAC function has to resist existential forgery, i.e., an attacker should not be able to forge any valid couple (m, σ) without the knowledge of the shared key K.

The MAC function can be built by either cipher block algorithm or cryptographic hash functions. A cryptographic hash function, also sometimes called a one-way hash function, has the following characteristics:

- The hash of any arbitrary size of data will always have a fixed size.
- Changing any bit of the hashed data generates a very different hash result.
- It is extremely difficult to find a message m such that its hash value is equal to an arbitrary value.
- It is extremely difficult to find two messages m_1 and m_2 such that their respective hash values are equal. This property is called collision resistance.

Currently, the most widely used hash function is SHA-1. However, several recent studies demonstrated that the robustness of SHA-1 is seriously weakening [75, 76]. In 2005, NIST launched a public challenge to design the successor of SHA-1. MD5 is another widely deployed hash function. MD5 hashes often serve as unique identifiers for torrents in P2P networks.

A MAC function that uses a cryptographic hash function is called a Hash-based Message Authentication Code. The most popular HMAC functions are HMAC-SHA1 and HMAC-MD5.

3.2.3 Asymmetric Cryptography

In 1976, two researchers, Whitfield Diffie and Martin Hellman, in a famous seminal paper [77] built the foundations of a new type of cryptosystem: asymmetric cryptography. Here, Bob owns a pair of keys: the public key K_{pub} and the private key K_{pri}. Both keys are linked with a mathematical relationship. To securely send a message m to Bob, Alice obtains a copy of Bob's public key K_{pub}. She applies to message m an algorithm E, called encryption, using Bob's public key K_{pub}. The result ciphertext is $c = E_{\{k_{pub}\}}(m)$.

To retrieve the clear text, Bob applies to message c an algorithm D, called decryption, using his private key K_{pri} such that $D_{\{K_{pri}\}}(c) = D_{\{K_{pri}\}}\left(E_{\{K_{pub}\}}(m)\right) = m$ (Fig. 3.3).

The encryption and decryption algorithms are designed such that, at least in theory, without the knowledge of the private key K_{pri}, it would be infeasible for Eve to find m, or, more precisely, any information about m except its length, from c (except by breaking the underlying hard to solve mathematical problem). Therefore, it is paramount that Bob keeps his private key secret. However, as its

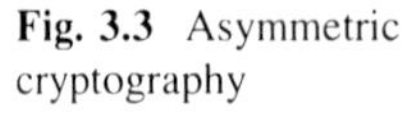

Fig. 3.3 Asymmetric cryptography

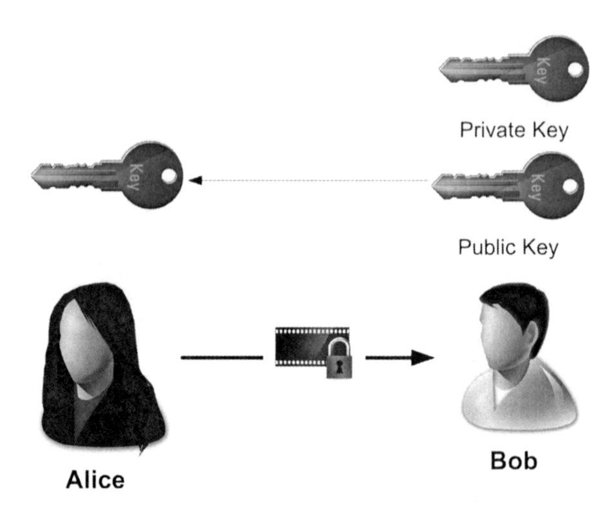

name indicates, Bob can safely distribute his public key to everybody, even to the malicious Eve.

Like MAC, asymmetric cryptography can provide both integrity and authentication. Bob wants to send a message m to Alice. He does not look for confidentiality but he wants Alice to be sure that the message is originated by him (authentication) and that Charlie did not tamper with the message (integrity). For that purpose, Bob signs the message m. Bob applies to m a mathematical function called signature using his private key K_{pri}. The result σ, called a signature, is $\sigma = S_{\{K_{pri}\}}(m)$. He sends together m and σ to Alice. Alice verifies the signature using Bob's public key K_{pub}. The signature σ is valid if $V_{\{K_{pub}\}}(\sigma) = m$, where V is an algorithm called Verification. In other words, the respective roles of the public and private keys are inverted. The main difference with MAC is that the signing entity and the verifying entity use different keys.

In fact, most implemented signature schemes use an additional cryptographic hash function. Using the cryptographic hash function H, we obtain the new signature scheme $\sigma = S_{\{K_{pri}\}}(H(m))$ and the signature σ is verified if $V_{\{K_{pub}\}}(\sigma) = H(m)$. Signing the hash of the message rather than the full message has several advantages such as a fixed length of signature, regardless of the size of signed data, and efficiency. Calculating the hash of a large dataset is faster than signing the same large dataset with asymmetric cryptography.

The robustness of asymmetric cryptography relies on the difficulty of solving some difficult mathematical problems such as the factorization of composite numbers made of two large prime numbers, or the difficulty of calculating discrete logarithms in finite fields or on elliptic curves.

Currently, the most widely used asymmetric cryptosystem is RSA. RSA stands for the initials of its three inventors: Ron Rivest, Adi Shamir, and Leonard Adleman. RSA can be used for both encryption and signature. It was invented in

1977 [78] and is still considered secure. RSA Data Security patented this algorithm. Since 2000, the algorithm has been in the public domain. It is currently estimated that RSA requires a key length of at least 1,024 bits to be safe [72]. The security level of a 1,024-bit key of RSA is considered equivalent to the security level of an 80-bit key for symmetric cryptography. NIST proposes to phase out the 1,024-bit keys in 2015 and to replace them with 2,048-bit keys (equivalent to a 110-bit key for symmetric cryptography).

3.2.4 Hybrid Cryptosystems

Symmetric cryptography is fast and suitable for encrypting large datasets, where as asymmetric cryptography is much slower. Symmetric cryptography is at least 100 times faster than asymmetric cryptography. Then, why should we use sloppy asymmetric cryptography?

The major reason is key management. Let us suppose that Alice and Bob only use symmetric cryptography (which would have been the case before 1976). If Alice wants to send a message to Bob, she has to use a key shared with Bob, K_{ALICE_BOB}. Now, if she wants to send a message to Charlie, she has to use another key that this time she shares with Charlie, $K_{ALICE\_CHARLIE}$. Obviously $K_{ALICE\_CHARLIE} \neq K_{ALICE\_BOB}$, since otherwise Bob (respectively Charlie) could eavesdrop on the conversation between Alice and Charlie (respectively Bob). If Bob wants to securely send a message to Charlie, they have to define their own common key: $K_{CHARLIE\_BOB}$. Obviously, the more people participating, the more keys needed. A population of n members would need to define $\frac{n \times (n-1)}{2}$ shared keys. As an example, 1,000 members would require 500,000 different keys. This would quickly become unmanageable.

Asymmetric cryptography simplifies this problem. Each member needs a unique public–private key pair. Each member keeps its private key secret and publishes its public key, for instance, in a certified directory of keys. Thus, if Alice wants to securely send a message to Charlie, she has just to obtain Charlie's public key from this directory. A population of n members would need to define n key pairs. The previous community of 1,000 members would only need 1,000 key pairs compared to the previous 500,000 symmetric shared keys. Unfortunately, if the message to encrypt is large, using an asymmetric cryptosystem may consume too much time. Fortunately, there is a solution.

Alice generates a random number, which we will call the session key K_s. Alice then encrypts it with Bob's public key K_{pub_Bob} and sends K_s to Bob. Only Bob can decrypt the encrypted session key using his private key K_{pri_Bob}. Now, both Alice and Bob share a common key K_s. They can use this session key to encrypt large messages with a symmetric algorithm without much time penalty.

In fact, the protocol is often more complex. In the previous protocol, Alice is sure that only Bob can read her message. Nevertheless, Bob cannot be sure that the message was indeed issued by Alice (or rather by Eve). He only knows that the emitter knew his public key, which is by definition common knowledge. Thus, the first modification is to apply a mutual authentication, i.e., both principals authenticate each other. Seemingly now, Bob knows that Alice is on the other end of the line, and vice versa. Unfortunately, this may be insufficient.

Let us introduce a typical attack, called the man-in-the-middle attack. Alice and Bob both have a public–private key pair (K_{pub_Alice}, K_{pri_Alice}) and (K_{pub_Bob}, K_{pri_Bob}). Normally, they should establish a secure channel using the previous protocol. Let us suppose that Eve has generated her own public–private key pair (K_{pub_Eve}, K_{pri_Eve}). First, she impersonates Bob by pretending to Alice that her public key K_{pub_Eve} is Bob's public key. She can now establish a secure communication with Alice. She does the same with Bob by impersonating Alice, i.e., pretending to Bob that her public key K_{pub_Eve} is Alice's public key. She can establish now a secure channel with Bob. She is in a perfect position, where as long as she forwards to Alice the messages she received from Bob and vice versa, she can stealthily eavesdrop on the supposedly protected communication. Alice and Bob will not be aware of the ongoing spying. Obviously, Eve may even alter the contents of the exchanged messages. The problem is that the public key is just a large, random number without any attached semantics. Thus, anybody can replace a public key with another one. Alice has no way of knowing whether the presented public key is really Bob's, unless of course if Bob handed it directly to her. Unfortunately, this is not convenient in most cases.

The solution is to link the public key to the identity of the principal owning the key pair. For that purpose, asymmetric cryptosystems use two new elements:

- a key certificate, and
- a Trusted Third Party that will act as a Certificate Authority (CA); obviously every member of the community trusts this authority.

The key certificate is a digital data structure that contains at least the following elements:

- the public key itself,
- the identity of the owner of the key in a human-readable form,
- the identity of the issuer of this certificate (i.e., the CA) in a human-readable form,
- the signature of the certificate by the CA, which by signing this certificate, certifies that the owner of the public key is the one identified in the certificate. Thus, when Bob receives the certificate of Alice's public key, he will verify its signature using the public key of the Certificate Authority, often called the root key. If the signature matches, then Bob knows that it is indeed Alice's public key. This of course implies that Bob trusts the Certificate Authority. This implies also that the CA verifies the real identity of Alice prior to delivering the certificate of her public key. Furthermore, now Eve cannot anymore impersonate Alice. If Eve forges a certificate, she cannot properly sign it because she does not have the private root key. Thus, the signature checking would fail. She

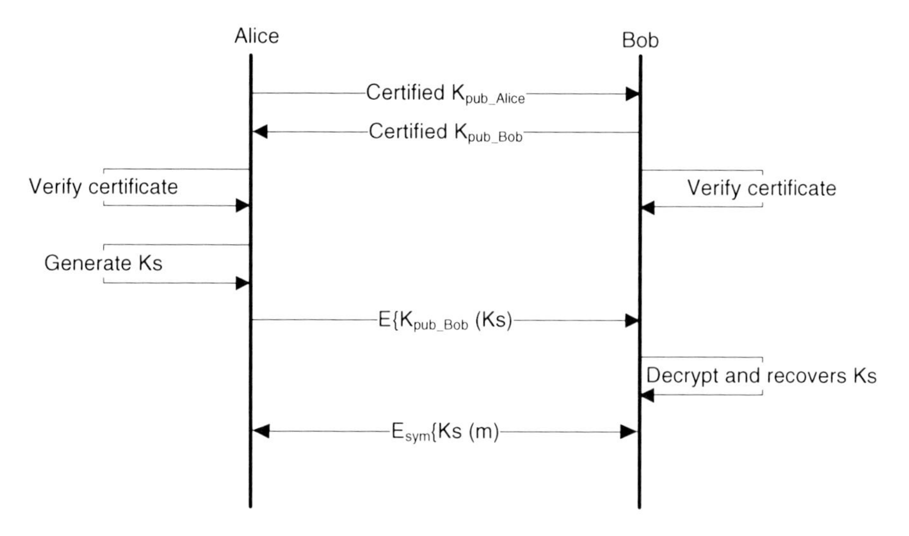

Fig. 3.4 Basic mutually authenticated exchange

cannot use a certificate that the CA would have delivered to her without disclosing her real identity. She cannot modify her certificate without spoiling its signature.

Today, most systems only use certified keys using trusted CAs. This prevents Eve from impersonating Alice and Bob. Figure 3.4 outlines the exchanges of a simplified authentication step. In this diagram, Alice does not prove to Bob that she knows the secret attached to her public key. For authentication to be complete, Alice would have to answer a challenge generated by Bob. At this point, the reader should guess that designing a secure protocol is an extremely complex task.

Certificates may contain many other fields such as validity period. In many cases, a key pair is only valid for a given period. When checking the validity of a certificate, it is mandatory to check that it has not expired. Due to the importance of certificates in modern cryptography, a standard for certificates was established in 1988: X.509 [79]. This standard has been updated several times (1997, 2000). Many widely deployed standards such as TLS/Secure Socket Layer (SSL), SMIME, or Pretty Good Privacy (PGP)[2] use the current version X.509 v3.

What happens if Alice loses her private key? For that purpose, we will use an additional element: Certificate Revocation List (CRL). The Certificate Revocation List contains the identification of all the revoked public keys and that should not be used anymore. The CA generates it. Thus, checking a certificate should require at least the following operations:

---

[2] In fact, PGP supports two formats of certificates: X509 and its own proprietary format.

- Verify the signature of the certificate.
- Check that the certificate has not expired if the expiration date is present in the certificate.
- Check that the CRL is the most recent one. If this is not the case, request the most recent one from the CA.
- Verify that the certificate has not been revoked.

Often, we refer to Public Key Infrastructure (PKI) as the infrastructure that manages the complete lifecycle of a public key. The Public Key Cryptography Standards (PKCS), published by RSA Laboratories, is a standard that defines many elements of a PKI [80]. Its use is widely generalized.

3.3 Digital Watermarking

3.3.1 The Theory

Digital watermarking consists of embedding an imperceptible message into a host multimedia signal without introducing perceptual distortions [81]. Watermarking hides a message *m* by slightly altering the original video frame F_o.[3] This is the role of the watermark embedder. The watermarking key enables us to pseudo-randomly hide the message, i.e., in a deterministic way that cannot be reproduced without the knowledge of this key.[4] The result is a modified video frame F_w that carries the message *m*. A single video frame may contain several messages, each using a different watermarking key.

To be invisible, modifications made to the images take into account the physiological characteristics of human vision. Let us give a crude example using the Least Significant Bit (LSB) of a pixel's luminance value. The LSB of the luminance value is set to 0 (respectively to 1) to embed a bit "0" (respectively a bit "1") in a particular pixel. The watermarking key, through a Pseudo-Random Binary Sequencer (PRBS), selects the pixels for embedding. The detector then locates the carrying pixels with the watermarking key and retrieves the hidden message by reading the LSBs of the luminance values. The human eye is basically insensitive to such small changes of luminance value. Nevertheless, it allows us to convey one bit of information per modified pixel. Of course, modern watermarks are more sophisticated.

The transmission of video frame F_w possibly introduces some noise. If the noise is acceptable, the consumer will perceive the received video frame as being

[3] We use video in this section. Nevertheless, the presented concepts are (most of the time) equally applicable to audio.

[4] Another interest in the secret key appears in legal cases. A court may demand disclosure of the algorithm used by the watermark. It will be part of the minutes of the process. Nevertheless, changing the key will allow us to safely continue using the watermark technology [82].

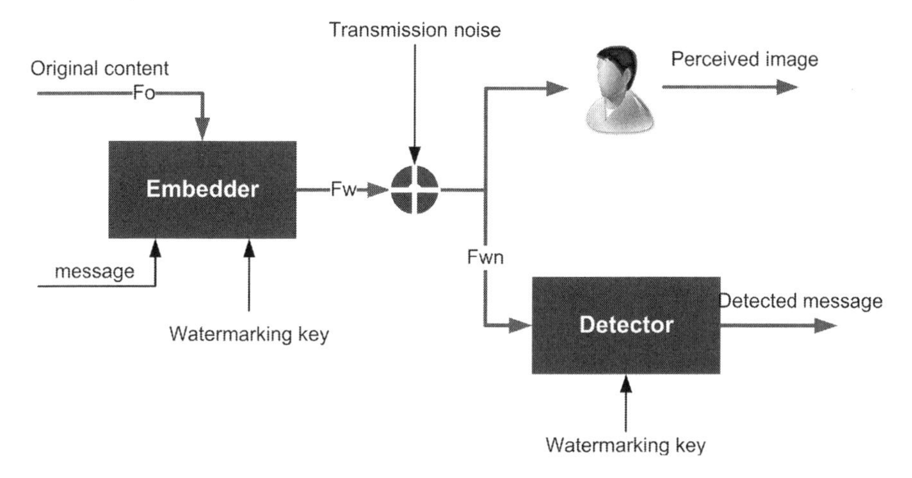

Fig. 3.5 Watermarking content

"identical" to the original one, even if it experienced slight alterations. In that case, the informed watermark detector, i.e., having the proper watermarking key, should still be able to extract hidden message m, i.e., watermark detection should be robust to transmission noise (Fig. 3.5).

- *Embedding rate* accounts for the amount of data that can be embedded in a unit of content. It is, for instance, expressed in bits per second or bits per frame for video. Useful information is often smaller than the number of embedded bits due to the use of Error Correcting Codes (ECC) such as Reed-Solomon [83]. Such error correcting codes are exploited to mitigate wrongly extracted bits due to transmission noise.
- *Robustness* refers to the ability to detect the embedded watermark even if the watermarked content experiences involuntary and/or voluntary transformations. Involuntary transformations are typically picked up during transmission along the video chain. They include filtering, lossy compression, digital-to-analog conversion, and desynchronization (resizing, rotation, frame rate conversion, etc.). On the other hand, voluntary transformations are generated by hostile attackers attempting to defeat the watermarking system. The attackers may have three goals: unauthorized embedding of a payload, unauthorized detection, and unauthorized removal of a payload. This last attack mainly consists of impairing the content so that the detector cannot anymore retrieve the payload. This type of attacks includes line removal, rotation, scaling, and camcording. Due to the targeted finetuning of such attacks, these voluntary transformations are usually for the detector harder to deal with than involuntary transformations. There is one additional type of voluntary attack: collusion attack. The attacker has access to differently watermarked copies of the same piece of content that she combines to alter or remove the watermark.

The classification as involuntary or voluntary transformation is by a semantic approach that attempts to model the goals of the "attacker". The classification is meaningless from a technical point of view since there is no difference between the two types of transformation for the detector. Thus, researchers often refer to attacks, regardless of their being voluntary or involuntary. Another possible classification of the attacks is under the categories of synchronous and asynchronous attacks. A synchronous attack does not alter the position of the watermarked samples. Lossy compression, filtering, contrast changing, black or white transformation are examples of synchronous attacks. An asynchronous attack alters the position of watermarked samples. Cropping, rotation, frame rate changing, bit rate changing, and camcording are examples of asynchronous attacks.

An attack is successful if it reaches its goal. For instance, in the case of removal attack, it hinders the detector from extracting the embedded payload, without degrading the perceptual quality of the video or audio beyond acceptable limits [84]. Acceptable limits define of course a very fuzzy frontier. Some benchmarks, such as StirMark for image watermarking [85], propose a relative reference framework for tests.

Digital watermarking is a complex tradeoff among these three parameters: fidelity, embedding rate, and robustness. Modifying one parameter necessarily influences the other two. Depending on the targeted use, the designer of the watermark will favor fidelity, embedding rate, or robustness.

Watermarking systems are also classified by the fact that the original non-watermarked content is required to perform detection.

- *Blind watermark* schemes do not need the original content to extract the payload information.
- *Non-blind watermark* schemes need the original content to extract the payload information. In some cases, having access to the original content may give an advantage to the detector in extracting the payload. This is especially true if the submitted sample has been badly impaired with asynchronous attacks. In that case, the operator in charge of extracting the watermark may manually perform some registration operations before actually performing the extraction. The goal is to bring C_{wn} as near as possible from C_O. The operator applies spatial and temporal registration, i.e., off-setting, desynchronization of frames, and geometrical distortion. The operator tries to mitigate the effects of the asynchronous attacks. However, depending on the targeted application, the original content may, in practice, not be available at the detector site.[5]
- *Semi-blind watermark* schemes require some metadata to extract the payload information. Nevertheless, these metadata do not enable reconstructing the

[5] Having the original content at hand is a constraint and a risk. You have to either completely trust the operators (which is possible if the operators are members of the content owner's company) or protect the original content. Thus, content owners prefer blind and semi-blind solutions.

original content. Ideally, the leakage of these metadata should not give a competitive advantage to the attackers.

It is recognized that blind watermarks are more technically challenging and complex to design than non-blind ones.

Watermark embedding and detection operations typically are performed in the baseband domain or some transform domain.

- Baseband domain means that the watermarking is directly applied to the uncompressed content. For video, it can reduce, for instance, to altering pixel values of the raw representation of the image or to slightly altering luminance. For audio, it can reduce to altering the phase of the audio signal. Early watermarking technologies mostly used baseband domains due to their simplicity and the low computational cost involved. Unfortunately, baseband domain watermarks provide poor robustness.
- Transform domain means that the embedding occurs after a (non-)linear transformation. Popular transforms include the Discrete Cosine Transform (DCT), the Discrete Fourier Transform (DFT) and the Discrete Wavelet Transform (DWT). Most modern watermarking techniques rely on such transformations as they provide useful spatio-frequency representations of the content, which allows controlling more easily the trade-off between fidelity and robustness. The main drawback is that these methods are also more computationally intensive.

One of the major advantages of digital watermarking is that the embedded watermark survives digital-to-analog conversion. Encryption protects only digital data. Nevertheless, in the end, digital video must always be rendered to the analog world, thus losing the protection of encryption. Thus, watermarks carry information, such as copy rules, copyright notices, and rendering device IDs, in both the analog and digital worlds.

3.3.2 Usage

Watermarks have many possible uses. Nevertheless, in this book, we will focus on the uses related to DRM and content protection. Rajan Samtani proposes three categories [86]: flag-based applications that enable copyright enforcement, forensics applications, and content identification-based applications. We propose a more usage-oriented taxonomy with four main usages [87, 88]:

- Forensic marking (similar to Samtani's forensics applications).
- Copy Control (similar to Samtani's flag-based applications).
- Copyright signaling (partly similar to Samtani's content identification-based applications).
- Monitoring, monetization (partly similar to Samtani's content identification-based applications).

3.3.3 *Forensic Marking*

The purpose of forensic marking is to pinpoint the source of potential content leakage in a distribution network. To do so, forensic marking needs two elements:

- A traitor tracing code that uniquely identifies the components of the network, e.g., users, and rendering devices.
- A binding technology that links the identifier to the actual piece of content.

Often, forensic marking uses digital watermarking to perform the binding operation with the "issuer".[6] The embedded message consists of a unique identifier that may identify the actual viewer of the content, the operator who processed it or the last principal that used it. This strategy is useful in several cases.

3.3.3.1 Movie Screeners

It is common practice to embed a watermark identifying the recipient of a screener sent to the members of the Academy of Motion Picture Arts and Science or of a piece of content sent to journalists for reviewing. Only the identified recipient should watch this content. The content of these screeners is usually not publicly available, making it extremely valuable to pirates. Leakage to the public domain may induce dramatic financial loss. In 2003, a work-in-progress copy of "The Hulk" appeared on P2P networks two weeks before its theatrical release [90]. This work print was not the final version. Many special effects were missing. Color adjustments were not finalized. As a result, early reviews on chat forums were scathing and the movie was unsuccessful.[7] With forensic marking, if the content were ever to be discovered in the public domain, for instance, on P2P file sharing networks, the embedded identifier could be extracted to identify the infringer [46]. This application requires the following characteristics:

- Fast Embedding: At the duplication stage, there is a need to make numerous individual copies of the same piece of content, each one carrying a different identifier. Therefore, the embedding process must be fast to avoid undue increase in production cost.
- Robustness to Collusion Attacks: Two or more recipients of the same content may collaborate to create a copy of the content that would carry none of their

[6] Forensic marking is sometimes referred to as fingerprinting, following the seminal work of Neal Wagner [89]. In 1983, Wagner proposed slightly modifying each copy of a digital content in an imperceptible way to discriminate between each instance of the same piece of content. Nevertheless, in this book, we reserve the term fingerprint for techniques of content identification; see Sect. 3.4. Other terminology is also sometimes used, such as traitor tracing, transactional watermark, content serialization, and user forensics.

[7] It is difficult to assess the real impact of this leakage to the movie's failure. However, the bad mouthing may have deterred many people from watching it in theaters.

own identifiers. This is an extremely difficult challenge that is not yet implemented in any current commercial solution [91, 92].

Forensic watermarking has been successfully used in many cases in this context. For instance, in 2004, the US actor Carmine Caridi, a member of the Oscar Academy, was fined $300,000 for sharing two award screeners ("Mystic River" and "The Last Samurai") with Russell Sprague. Sprague subsequently posted them on P2P sites [93]. Academy members sign a usage agreement that stipulates that they should neither "allow the screeners to circulate outside their residence or office" nor "allow them to be reproduced in any fashion, and not to sell them or give them away". Forensic evidence proved that Caridi violated this agreement.

3.3.3.2 In-House Tracing

In postproduction houses, tracking the processing path of a movie by embedding successive watermarks at different steps of the production/postproduction process is advised [47, 94]. It is thus possible, in case of leakage, to identify the last operation or even the last operator. This application requires the following characteristics:

- Fidelity: postproduction handles high quality video. This quality must be preserved.
- A way to embed successive watermarks and retrieve all of them. A piece of content may be processed several times before being finalized. Thus, successive watermarks can carry the complete history of the operations applied to the work.
- Robustness to typical postproduction transformations such as edits, colorimetric changes, format conversions, and geometrical transformations.

3.3.3.3 Digital Cinema

The Digital Cinema Initiative (DCI) recommends embedding a forensics mark just before projection [95]. The embedded message carries the time, date, video server identity, and location of the projection. It is thus possible to trace back to the theater where the camcorder capture occurred.

Furthermore, some digital techniques estimate the position of the camcorder within the theater by analyzing the geometrical distortion of the camcorded copy [96]. From this geometrical distortion, it is possible to extract an affine model that evaluates the solid angle of the axis of the camcorder. The watermark of the captured movie discloses the theater where the capture occurred. Given the geometry of the theater, it is possible to guess the seat by intersecting the axis of the camcorder with the actual geometry of the room. In particular, it is interesting to learn whether the capture occurred in the seating area of the theater or in the projection booth. The latter case implies complicity of the projectionist.

This application requires the following characteristics:

- Robustness to Camcording: In fact, in this scenario, the main threat is the illegal capture of video through camcording. The digital content itself is protected by scrambling, and is thus extremely difficult to access directly in digital form (Chap. 10, Digital Cinema). The weakest link remains the analog capture through camcorder.
- Real-time Embedding: The payload contains a real-time timestamp. Precise time information is indeed needed to discriminate between captures done during regular activity hours and captures done out of regular hours, which requires complicity of the projectionist.
- Fidelity: One of the promises of digital cinema is superior video quality. Thus, it is unacceptable that a watermark impairs it.

In India, E-City Digital banned Chetna Cinema from receiving new digital cinema movies because Chetna Cinema made no attempts to prevent camcording in their premises [97]. Forensic watermarks proved that bootlegs were coming from camcorder captures in their theater.

3.3.3.4 Business-to-Consumer

There is currently an emerging trend to watermark the video output by consumer electronics devices [98]. A unique forensic watermark identifies either the rendering device or the purchaser. There are two potential architectures [99]:

1. Watermark embedding occurs on the content server side. The unique payload is defined by the license server and the content server embeds the watermark before scrambling the content. This solution has the advantage of being transparent to the client application. From the point of view of security, it is optimal under the reasonable assumption that the server is more secure than the client application. Nevertheless, this solution introduces real-time and computational constraints on the content server since it uniquely watermarks the content for each customer and encrypts this watermarked content. Each watermarked content being unique, it is impossible to use multicast mode; only unicast mode is possible. Thus, transmitter-side forensic marking is both computational and bandwidth demanding.

Of course, this approach is not compatible with the pre-scrambled VOD approach. In the pre-scrambled approach, the pieces of content are already encrypted on the file server. The same scrambled content is streamed to every consumer for optimization. Therefore, adding watermarking would be counter-productive as it would require descrambling prior to watermark embedding. A mitigating strategy could restrict the watermarking to some parts of the content. These parts would be stored in the clear on the file server. This would only require partial real-time scrambling.

Another possible approach is to select some parts of the content and for each part prepare copies that each carry a different payload. All the elements can be pre-scrambled. When sending the content to a user, the content server selects the right version of each watermark part. A variant of this approach is used in Blu-ray discs (Sect. 8.5.2, Sequence Keys).

2. Watermark embedding occurs at the application client side. At rendering,[8] the unique payload may be defined by the client application itself or by the license server. In the first case, the application has a unique identifier (or may use a unique identifier of its hosting device). In the second case, the payload is securely carried in a license. This solution drastically reduces constraints on the transmitting server. Unfortunately, its security is also lower. In most cases, the attacker has access to the client application. If it is not well protected (Sect. 3.6, Software Tamper-Resistance), the attacker may defeat the system. The attacker has two possible strategies. She may "simply" bypass the watermarking process and issue watermark-free content. Alternatively, she may modify the payload. In that case, the content may trace back to an invalid identity, or, more critically, frame an innocent user.

Embedding at the client side has several advantages, among which are the facts that it supports multicasting and is far less constraining than the server side. Nevertheless, it is less resistant to hostile adversaries. Thus, it may face some problems regarding non-repudiability in court.

The commercial value of multimedia content protected by forensic marking is extremely high, especially for so-called screeners used before the official release. Thus, these pieces of content are attractive targets for organized pirates. Fortunately, the trust model of forensic marking is relatively strong. It assumes that the average pirate has access neither to the embedder nor to the detector. Only employees in charge of marking the screeners should normally have access to the embedder. Moreover, only trusted employees should have access to the detector for investigation purpose. Therefore, it is reasonable to assume that the pirate will operate blindly. The objective of the pirate is either to remove the embedded watermark to avoid being traced or modify the embedded watermark so that an innocent is traced, or to void it. If a pirate applies a transformation to a piece of content, she has no means to assess whether the attack was successful or not. In other words, she is not certain of not being traced. Thus, to be relatively safe, the pirate has to make extremely strong modifications that severely degrade the quality of the piece of content, making it less valuable.

---

[8] Indeed, depending on the type of devices, there are other possible positions for the embedder than prior to the rendering process. If the watermark technology operates in the compressed domain (or at least partly), in the case of PVR, the piece of content would be watermarked prior to being recorded on the internal hard drive. Similarly, in the case of a home network media gateway, the piece of content would be watermarked prior to being ingested by the home network [100].

While the main assumption is that the pirate has no access to the watermark detector, two factors can invalidate this assumption:

- The pirate is part of an organization that gained, by some means, access to such a detector (typically class III attackers). This demonstrates how paramount it is to control who accesses the detector and who has access to the secret keys used by the embedding/detection process.
- The same watermarking technology is used for a second purpose such as copy control or copyright control. In that case, the pirate gains indirect access to the detector. Thus, it is recommended that the watermarking technology used for forensic marking applications are only used for that purpose. Other usages should employ a different watermarking technology.[9]

3.3.4 Copy Control

Copy control uses the embedded watermark to encode information defining the authorized usages of the marked content. For instance, the embedded message may be the so-called Copy Control Information (CCI) defined by the Copy Generation Management System (CGMS). It defines four states:

- Copy never: there should be no copy of this piece of content.
- Copy once: one generation of copy is allowed, but not more.
- Copy no more: it is the result of the copy of a "copy once" marked content. It should not be further duplicated.
- Copy free: the piece of content can be copied without any restriction.

When a compliant recorder receives content to duplicate, it should check for the presence of a copy control watermark. If any is found, it should enforce the prescribed policy as follows.

- If the content is marked "copy never", then the recorder has to cancel the recording operation.
- If the content is marked "copy once", then the recorder can make the copy. The watermark carried by this newly generated copy should be changed to "copy no more". This is the trickiest part. It means that the recorder needs a watermark embedder to insert this mark.[10] Thus, copy control offers the pirates access to both a limited embedder and a detector. This double feature weakens the trust model.[11] With the embedder, the attacker has the piece of content before

---

[9] Even if they use different keys, the trust model would be weakened. .

[10] Most proposed solutions did not erase the "copy once" mark. Therefore, a "copy no more" content holds both "copy once" and "copy no more" marks.

[11] Some implementations proposed using a different watermarking technology to embed the "copy no more" mark. This solution does not enhance the trust model because the detector needs still to handle both watermarking technologies.

watermarking and the same piece of content watermarked with a known embedded message. This is precious to the attackers, who can compare the two pieces of content to locate watermarks. Furthermore, the detector provides a very handy tool to the attacker. She can check if she succeeded in defeating the watermarking system.
- If the content is marked "copy no more", then the recorder has to cancel the recording operation.
- If the content is marked "copy free", then the recorder can make the copy. It does not have to change or add new watermark.

Watermark detection is performed for both digital inputs and analog inputs. This obligation, as well as the mandatory behavior, is defined and enforced by the compliance and robustness regime (Sect. 3.8, Compliance and Robustness Rules).

Although the first design started in 1997 [101], CGMS has not yet been implemented, nor been fully standardized. The problems are not any more technical, but rather economic and political.

3.3.5 Copyright Management

Copyright management uses digital watermarking to convey two types of information:

- Information related to copyright such as name of the issuer, name of the rights owner, and unique identifier of the piece of content.
- The fact that the content is protected by copyright and thus may have to obey some rules.

Historically, the first application foreseen for watermark was copyright management [102]. At that time, restricting copying through DRM was perceived as a bad solution. Nevertheless, if it were possible to prove potential infringement of copyrighted materials, then there would be no need for restrictions. The aim of digital watermarking was to provide a means for such a proof by inserting copyright information that is invisible, unalterable, non-removable, and robust to any modifications. Furthermore, as with forensic marking, it was believed that the fear of litigation due to this marking would be a deterrent to illicit duplication. History proved that this belief was wrong.

Copyright watermarks also provide a way to monitor the diffusion of content in the broadcast environment [103]. They can verify whether a program is not broadcasted more often than agreed or rebroadcasted by another broadcaster without authorization.

Compared to forensic marking, its main difference is the carried payload. In forensic marking, the watermark traces back to the user, whereas in copyright control, the watermark traces back to the original author or to the rights holders.

This type of usage requires a trusted authority. In the most common schemes, the roles of the trusted authority are the following:

- To register the copyrighted work. The authority has to check that the submitted work is genuine and not already registered. Then, it defines a unique identifier for this piece content. It logs the unique identifier, the identity of the requester, the embedded payload, the time and date of registration, and characteristics of the content. Furthermore, the trusted authority deposits the work in its archive.
- To embed a watermark in the content that uniquely identifies the work using the payload it previously calculated.
- To offer a service that extracts watermarks, identifies the associated work in its database, and feeds back the corresponding information such as the name of the work, the identity of the rights holder, and the date of registration. Anybody should be able to query this information.

There are many variants of this scheme. For instance, the rights holder may locally watermark his content with his own payload and then send the already watermarked content to the trusted authority for registration. The trusted authority will serve as an arbiter in the event of a dispute about ownership.

Proving the ownership of content requires three steps:

- Extract the payload embedded in the examined piece of content. This operation requires knowledge of the watermarking technology and the secret key used for the watermark embedding process.
- Verify that the embedded payload corresponds to the registered watermark. In other words, the payload being unique per opus, the watermarked content should be the same as the registered opus present in its archive.
- Resolve potential conflicts. In practice, a piece of content may carry several copyright watermarks. They may all come from the same trusted authority or they may come from different authorities. In some scenarios, this may be normal. Imagine that the creator of one piece of content includes an excerpt of another piece of content. Then, the excerpt may already have its own watermark identifying its original author. The new piece of content may legally carry two watermarks (at least in some parts): one from the original opus and one from the newly created opus. In the case of litigation, the piece of content may hold two watermarks. The time of registration will help resolve eventual conflicts. It corresponds to the time of watermark embedding. This determines, in theory, the first embedded watermark, i.e., the legal owner of the opus. This way of resolving conflicts using time racing conditions does not require using a single trust authority as long as they are all trusted and synchronized.

One of the issues of watermarking technology is its reversibility. If an entity can extract the payload, in theory, it has also the ability to erase this payload, or to forge a new payload. In fact, to read a watermark, an entity needs to know both the watermarking technology and the secret key used for embedding. In other words, the reader has all the needed information to embed a watermark. To deal with reversibility, cryptography complements watermarking [104]. Using public key cryptosystems to encrypt the payload allows breaking the reversibility at the system level. The owner of the content will also have the private key used to encrypt the embedded payload. New schemes of copy protection interlace cryptography and watermarking.

Furthermore, some schemes combine two uses of watermark. For instance, Alessandro Piva [104] proposes a scheme using three successive watermarks. The first watermark holds the identity of the issuer of the content. In other words, the first watermark holds the copyright information. The second watermark holds the identity of the distributor, together with the usage rights. The author applies this second watermark to prove that it granted the rights to distribute the opus. This watermark serves the need of usage control. The third watermark holds the identity of the customer who purchased the opus from the distributor. This watermark serves the need of forensics.

3.3.6 Monitoring, Monetization

Broadcast monitoring basically consists of surveying the consumption of a piece of content [105]. This can be exploited, for instance, to measure the audience of TV and radio channels. The channels that want to measure their audience embed a digital watermark in their broadcast program. The payload may carry a content ID, a channel ID, a distribution network ID, and the airing time. Volunteer panelists have a box containing a watermark detector. The box automatically extracts the watermark and sends a detailed report to a central collecting server. It is then easy to infer the audience from this panel minute by minute. By comparing the embedded airing time with the captured time, it is even possible to detect time-shifted viewing, i.e., when people record an event and watch it later. Since 2008, all French digital terrestrial channels have embedded an audio watermark, which permits the French company Médiamétrie to measure audience [106].

Broadcast monitoring infrastructures can also serve as simplified forms of forensics [107]. By monitoring potential venues of illegal content, it is possible to quickly react to any illegal rebroadcast in real time. It is thus possible to detect in a few minutes the appearance on a UGC site of pirated broadcast content. This is especially true for special live events such as the Super Bowl, title fights, the Olympic Games, and presidential debates. These live events offer interesting revenue opportunities. In a few seconds, it is possible to send a takedown notice for the offending content.

3.4 Fingerprinting

This type of technology is not strictly speaking a protection technology. Nevertheless, it is sometimes part of a larger framework of protection. It is a standard tool in the fourth security goal presented in the introduction: scouting.

Fingerprinting works in two steps. The first step is the registration of the original content from which the process extracts all the fingerprints. These fingerprints are stored in a database that we will call the reference database. Fingerprinting technology can identify a piece of content only if it has prior knowledge of the original content.[12] The second step is the identification of candidate sample. It requires several operations. Often, the first operation normalizes the candidate sample, i.e., translates into a standardized temporal and spatial format similar to the one used by the references. The second operation extracts the fingerprints as was done for the original content. The third operation submits the extracted fingerprints to the reference database. The answer does not look for a perfect matching as in a conventional relational database. It rather provides a number of candidates with strong similarity. The similarity is measured by the distance between the submitted fingerprints and the reference fingerprints. Last, a decision engine analyzes the returned candidates and takes its decision using multiple parameters such as the distance and the consistency of recommendations between successive images.

Fingerprinting technology extracts discriminating features from a piece of content, so-called fingerprints or multimedia DNA. There are mainly four types of image fingerprinting techniques [108]:

- Semantic fingerprinting acts at a high semantic level [109].
- Global approach-based signal processing analyses the entire image and extracts global descriptors such as the color/luminance histogram, texture descriptors, shape descriptors, and motion descriptors. This method has the advantage of being rather fast with low complexity but does not resist some attacks such as cropping and addition. It works mainly for "soft" attacks.
- Local approach-based fingerprinting extracts salient or key points of the image. Key points have the advantages of being independent, being precisely localized in a frame, and carrying discriminating information. Thus, key points are extremely valuable for fingerprinting. Each descriptor models the neighborhood of one key point. This method is extremely robust to many transformations, and even to camcording. It has the disadvantage of being slow and of generating larger fingerprints [110].
- Perceptual hashing has similarity with cryptographic hashing in the sense that it is easy to compute, presents weak collision risk, and is difficult to reverse (i.e., it

[12] In the operating method, video and audio fingerprints are similar to the human fingerprints used by police. The fingerprints must be in the central database so that the police can potentially identify a suspect. As in video, although the submitted fingerprints may be altered (partially in the case of human fingerprints), the police may still be able to identify the suspect.

is computationally difficult to forge an image that generates a given hash. It is referred to as one-way function). The main difference is that a small modification of an image should generate a perceptual hash quite similar to the perceptual hash of the original image [111]. The similarity is measured, for instance, by Hamming distances.

Fingerprinting technology is complex and requires a delicate trade-off among accuracy, speed, size of the fingerprints, and robustness. Depending on the targeted applications, the fingerprinting technology should match the transformed content with the original content. The transformations range from cropping, resizing, rotating, and compressing to camcording.

3.5 Hardware Tamper-Resistance

3.5.1 Definition

Hardware tamper-resistance is the set of technologies that make a hardware component such as a processor or memory resistant to physical and logical attacks. As is usual in security, this type of definition offers a large set of possible interpretations. We may be more precise using a definition from Keith Mayes [112]. A secure component has the following characteristics:

- It participates in an electronic transaction; this eliminates all the techniques used just to seal a device (although they may be used as additional protection, as in a Hardware Security Module (HSM) when detecting the case being opened).
- The primary purpose of this component is security.
- The component is difficult to reproduce and especially difficult to modify.
- The component stores data in a secure way, i.e., it should be extremely difficult to access the data in the clear by either logical methods such as calling APIs or physical methods such as microprobing.
- The component executes security-oriented functions, for instance, crypto-graphic algorithms. They even sometimes implement cryptographic co-processors. Cryptography is highly demanding in computing power. Co-processors may help accelerate calculations through wired implementations of algorithms, such as exponentiation. Thus, a co-processor with exponentiation acceleration performs a RSA-1024 signature in a few milliseconds, whereas a generic processor would need several seconds.

From this detailed definition, it is possible to derive some security characteristics:

- The secure component acts as a safe for data, i.e., there is no way to get illicit access to the stored data.

- The secure component has trusted behavior, i.e., it is not possible to modify its executing software.
- It is possible to have a unique instance of a secure component using unique stored data. This instance cannot be easily cloned.

Obviously, these characteristics are extremely valuable for building a trusted environment, which is paramount in content protection.

3.5.2 Different Types of Components

There are mainly four types of components that use tamper-resistant hardware:

- Secure processors are processors whose unique purpose is to perform security-related operations. They come in two main types of form factors:
  - Packaged chip to be soldered on printing boards.
  - Packaged chip to be embossed into smart cards, i.e., plastic cards.
- Hardware Security Modules are highly specialized devices mainly dedicated to key management. They come as PC extension cards or rackable units.
- Systems on Chips (SoC) are large chips that assemble, on the same piece of silicon, different physical components to provide a larger set of features. Security is not the main purpose of SoC but often an expected feature. Thus, it does not comply with the second requirement of the proposed definition. Nevertheless, SoCs are widely used in Set Top Boxes and gateways and implement many security features for Conditional Access Systems.
- Trusted Platform Modules (TPM) are secure processors whose unique purpose is to build a trusted environment for a system

3.5.2.1 Secure Processors

Secure processors, especially in the form of smart cards, are probably the hardware tamper-resistant components most familiar to consumers. Consumers may use smart cards every day. In 1974, with his first patent of a very large portfolio of patents, Roland Moreno defined the foundations of the future smart cards [113]. It was many years before Bull proposed the first commercial smart card [114]. Pay TV, with Eurocrypt and Videocrypt, was one of the first consumer applications using smart cards (Sect. 6.2, Pay TV: the Ancestor). Then, in 1986, the French GIE Carte Bleue, a consortium of banks, decided to use smart cards for the Carte Bleue credit cards. The smart card still carried the usual magnetic strip, but the preferred transaction mode used the secure component rather than the magnetic strip. Since 1993, every French credit card has been a smart card. Following this success, more credit card associations switched to smart cards, ending up with the current international standard EMV (Europay, MasterCard, Visa). However, the big win for the smart card industry was its adoption by the European standard Global System for Mobile communications (GSM) as the security agent of mobile phones. GSM required a smart card, so-called the Subscriber Identity Module (SIM), for authentication, encryption, and additional services. Smart cards are used in other areas such as travel and access control to buildings. Many new passports embed also a contactless smart card.

Smart cards are of mainly two types:

- Contact cards connect to a terminal through a standardized eight-pin connector defined in the ISO 7816 standard [115]. The terminal or card reader provides power supply and clock signals to the component. There are mainly two form factors: the credit card size and the SIM form factor, which is smaller and is in use in mobile phones. In fact, the secure processor itself is far smaller. It is only the dimension of the plastic package and the precise position of the eight-pin connector that are normalized.
- Contactless cards do not need a physical connection to the terminal. The power supply, the clock, and the exchanged data are delivered wirelessly. There are three standards, depending on the frequency of the carrier signal.
  - Close-coupled cards operate at a very short range of less than 1 cm. The transfer rate is low: 9.6 Kbits/s. They use a carrier at 307.2 kHz. They are defined by the standard ISO-IEC 10536 [116].
  - Proximity cards operate in the range of 20 cm. The minimal transfer rate is 106 Kbits/s and they can reach up to 817 Kbits/s. They use a carrier at 13.56 MHz. They are defined by the standard ISO-IEC 14443 [117]. These are the most widely deployed cards. They are present in e-passports, transport ticketing and contactless banking cards. The provided power supply is sufficient for using cryptographic co-processors.
  - Vicinity cards operate in the range of 10–70 cm. The transfer rate is a few Kbits/s. They use a carrier at 120 kHz. They are defined by the standard ISO-IEC 15693 [118].

Smart cards due to their usage in consumer domain have to be "cheap". Nevertheless, the field of applications requires having a serious level of security. They are an interesting target for pirates due to their wide usage and market. Furthermore, being mobile devices, they must have limited power consumption.[13] Size and power consumption limitations (implying also slow clocks) lead to reduced computing power. In fact, smart cards are far less powerful in terms of performance than the usual processors used in computers.

[13] Low consumption constraints did not initially come from the fact that smart cards were removable. Indeed, for the first foreseen applications, the smart card reader could have provided enough power supply. In fact, the reduced consumption constraint derives from the small form factor and the associated problems of thermal dissipation. The more power consumed, the more thermal energy produced. Too much heat would destroy the component. Later, with its utilization by mobile devices such as mobile phones, or for wireless cards, the power consumption became a crucial issue.

3.5.2.2 Hardware Security Modules

Some applications require intensive computing power that smart cards could never provide. This is the role of Hardware Security Modules (HSM). HSMs are extremely powerful calculators that are dedicated to cryptographic applications. They have two main characteristics:

- Dedicated, integrated cryptographic co-processors, which perform exponentiation, drastically accelerate public key cryptography.
- Extremely secure components: like smart cards, they include many tamper-resistant features. For instance, a HSM may physically fully erase its secure memory once it detects any strange behavior by the power supply (for instance, to prevent fault injection analysis). It securely erases its secure memory if the cover is ever removed.

The predilection domain of a HSM is the management of PKI. The root keys of any hierarchical key structure should be stored in HSM. The assumption that the root key does not leak is probably the most important security assumption of any trust model using public key cryptography. Therefore, the HSM provides the expected high level of security. PKI requires signing certificates. PKCS, the major standard, often uses RSA-1024 or, even better, RSA-2048. If a large number of certificates has to be generated, the signature phase is a very lengthy operation. Therefore, the HSM offers the capacity of its dedicated accelerated cryptographic co-processor. Of course, the private keys never leave the HSM. The only exception is for backup. When initializing a HSM as a root key, during a so-called key ceremony, the backup transfers the private key(s) to another HSM or to smart cards using secure protocols. These backup HSMs are then securely stored in disseminated locations (to survive physical catastrophes).

Another domain of predilection of a HSM is the acceleration of secure connections. For instance, a HSM can quickly perform the initial handshake of SSL that uses asymmetric cryptography.

A HSM should have at least the following:

- A FIPS 140 validation: the FIPS 140-2 standard defines security requirements for a HSM [119]. Indeed, the FIPS 140-2 standard defines four levels of security.
  - Level 1 is the lowest security level. It does not require any tamper evidence. It is mostly for software products or production grade components.
  - Level 2 requires some form of tamper evidence.
  - Level 3 adds additional requirements such as physical or logical separation between interfaces handling critical information and identity-based authentication, erasing the secrets when tampered, and the audit verifies that the implementation of the cryptographic algorithms is state of the art. Nevertheless, it can still be defeated by skilled attackers of class II of IBM's classification [120]. Thus, these levels should be avoided for any application requiring serious security or operating in a hostile environment. The gap between level 3 and level 4 is huge.

  - Level 4 of FIPS 140-2 targets attackers of class III of IBM's classification, i.e., funded organization. It requires robustness against environmental attacks. For instance, the certification of this fourth level requires examination of the formal model of the chip and its detailed description. The evaluators will apply white box evaluation to certify the HSM.[14] The first HSM that received its FIPS 140-2 level 4 certification was the IBM 4758 PCI in 1998 [121]. Today, there are several commercially available level 4 certified HSMs [122] and many at FIPS140-2 level 3.

- A real random number generator: only a true source of entropy can generate true randomness, which is key for cryptography. A source of entropy is the measurement of physical noise such as thermal noise.
- Secure clock, i.e., an autonomous clock that cannot be altered by external commands or events.

HSMs are expensive tools but powerful and extremely secure components, especially at level 4. The cost of HSM ranges from a thousand dollars to several tens of thousands of dollars for the most powerful ones.

3.5.2.3 System on Chips

SoCs are components that integrate many non-security-related important functionalities in one unique physical component. Modern Set Top Boxes (STB) use such chips that integrates a generic main processor (such as ARM, ST, and Atom), a video processor that decodes and renders video signals, and an audio processor. Some chips integrate dedicated video descramblers that support CAS. In that case, the video path and the descrambler are in a tamper-resistant area together with a secure storage area. The main processor has no access to this tamper-resistant area. This architecture offers a higher level of security. The video is never available in the clear on any external bus. Furthermore, often the video memory is encrypted, avoiding simple dumping attacks.

SoCs are used in most domains using embedded processors. Of course, they integrate other functionalities and do not necessarily embed any security features.

3.5.2.4 Trusted Platform Modules

Trusted Platform Modules are secure components compliant with the standard defined by the Trusted Consortium Group (TCG) [123]. The objective of a TPM is

[14] Evaluating a security system in white box means that the evaluator has access to the documentation of the internal behavior of the system. Evaluating in black box means that the evaluator has only access to public information. He may only guess the behavior from what he publicly learns. Therefore, white box evaluation is more challenging for the designers than black box. In white box, the evaluator acts like an insider.

to establish a root of trust. The idea comes from a paper published in 1997 on a reliable and secure bootstrap architecture [124]. When the computer boots up, the following process occurs:

- The secure processor checks the integrity of its boot ROM, executes the boot ROM, and checks the state of the machine. If successful, it goes to the next step.
- Then it checks the integrity of the kernel of the Operating System (OS), executes it, and checks the state of the machine. If successful, it goes to the next step.
- Then it checks the integrity of the hardware and the integrity of corresponding drivers and executes them and checks the state of the machine. If successful, it goes to the next step.
- Through an iterative process, the secure processor expands the boundary of the trusted hardware and software.

In fact, the secure processor does not check for integrity. It securely stores signatures and seals secret data for a given state of the machine. Each element of the trust chain stores the signature of the next step and verifies that the sealed data is correct.

Once the machine is in its final state, the secure processor may certify the configuration of the machine to eventual third parties. Integrity checking means extracting a cryptographic hash of binary code, data values in registers, and some characteristics of the physical components of the system. The secure processor performs the hash calculation as well as the comparisons. If the comparison matches, the system is trusted. The TPM would spot any modifications of the hardware of the system as well as of the software. In addition, TPMs provide secure storage for keys and certificates.

Many recent computers already embed a TPM. Nevertheless, the software executing on TPM-equipped computers rarely uses them. Unfortunately, no widely deployed operating system supports its usage or even uses it. In 2002, Palladium, an early initiative of Microsoft, made headlines. Its detractors presented Palladium as a spy within the computer that would control what could and could not be executed on the computer [125]. Palladium received a very negative response from the public. This initial rebuke slowed down the deployment of TPMs in the consumer domain. Microsoft delayed the design of a new generation of the operating system, code-named Long Horn that would have benefited from the TPM.

3.5.3 Dedicated Countermeasures

As specified in the initial definition, the hardware tamper-resistant processor component must be difficult to reproduce and especially difficult to modify and should store data in a secure way. For that purpose, the component embeds dedicated hardware features. We will give a non-exhaustive list of some of the deployed protected features.

The first type of protection attempts to counter physical intrusion. The attacker may try to access internal information by physically observing this information. This means either observing the value of a memory cell or spying on its value on a digital bus in the component. The usual way is to use microprobes or Focused Ion Beams (FIB) to measure electrical signals within the chip. A microprobe is an instrument using microscopic needles to measure voltage. A FIB can take a picture of a chip, remove some layers, and even change the value of a memory cell. Of course, these attacks require deep knowledge of microelectronics and electrical engineering. The attacker has to locate a transistor within a complex layout. The countermeasure must attempt to deter intrusion or detect intrusion. For instance, a lattice of overlaying metal layers may detect rupture of connection. It is a surface mesh with entangled ground, power, and sensor lines. This may happen if a microdrill attempts to drill through and break one of the lines. Nevertheless, a FIB workstation may go through this countermeasure.[15] Light detectors block the processor if ever light is seen, i.e., in a condition that normally should never be encountered. Indeed, seeing light means that several layers have been removed to directly access the architecture of the component. Often, there is also a detector of depassivation. Typically, on top of the circuit there is a thin layer of polymorph silicon called passivation layer. This is the first layer to be removed. For instance, this can be done by dropping a little bit of fuming nitric acid on it. The detector attempts to thwart this first attack.

As usual, serious and knowledgeable attackers will go through all these defenses. The second type of protection attempts to make reverse-engineering more difficult. The first countermeasure is to avoid regular patterns in the layout. The physical layout of a component presents many easily identifiable patterns such as memory cells, and digital buses [126]. These patterns help the attacker find its way in the layout. Randomness makes the chip less readable for the attacker. Some chips encrypt the communication bus. Thus, microprobing the data bus is useless. The attacker will have to look for other points of attack. Another useful precaution is to limit the reuse of the same component or layout. For instance, DirecTV used a solution already deployed and hacked in Europe. In addition, the solution was a variant of the European implementation; lessons learned from European crackers largely helped the US crackers (see "The Devil's in the Details" in Sect. 6.4).

The third type of protection attempts to detect unusual conditions of the environment. It may be interesting for the attacker to put the processor in extreme environmental conditions. For instance, when the temperature drops below minus 20 °C, the remanence of static Random Access Memory (RAM) increases.

[15] The newest techniques attack the chip from its backside with a FIB microscope, thus bypassing the sensor mesh.

Remanence of memory means that even after the power supply turned off, the state of memory remains valid for a given period. This means that the attacker may access RAM even once power supply off. Furthermore, under certain temperature conditions, writing to memory or reading from memory may be disabled. Many attacks may use this characteristic. Thus, secure processors often embed temperature sensors. Another typical attack is to execute the program very slowly, even step-by-step. In some cases, it allows us to get information through voltage surface analysis. Thus, secure processors often embed slow clock detectors. The same is true for other parameters such as the clock and the power supply. Glitches in the clock or the power supply are often the vectors of fault injection attacks. Variations in the power supply may provoke the processor to misinterpret or skip an instruction (for instance a conditional branching). Variations in the clock may provoke the processor to misread data or an instruction [127].

The fourth type of protection attempts to defeat a new type of noninvasive attack: the side channel attack. In 1999, Paul Kocher disclosed a devastating attack, which extracted secret keys through simple observation of timing [128]. The basic idea is rather straightforward. Let us suppose that the time to process an algorithm when one bit of the key is "0" is different from the time to process the same algorithm when this bit of the key is "1". Thus, statistical analysis with chosen ciphertext should be able to guess (or at least reduce the key space to explore). As an example, if the implementation does not perform an exponentiation when the bit of the key is "0", then there will be a clear timing difference with when the bit is "1" and an exponentiation occurs. Of course, this example is oversimplified. This attack was called a Simple Timing Attack (STA). Kocher and his team designed a more complex attack: the Differential Timing Attack (DTA). DTA extracted secret keys from most of the commercially available smart cards (at least at that time). Soon, the same principle of attack was applied to power consumption. The number of transistors commuting for a "0" may be different from the number of transistors commuting for a "1". This generated the Simple Power Attack (SPA) and later the more complex and more powerful Differential Power Attack (DPA).

The answers to this type of attack are both hardware-based and software-based. The countermeasure needs to smooth or randomize the timing and consumption all over the algorithm. This requires making dummy operations, adding some balance in the hardware architecture, and a salt of randomness. Side channel attacks are one of the most active areas of cryptography and secure IC design. In cryptography, as usual, the results are public. The algorithms used are not changed, but their actual implementations are fine-tuned to prevent these attacks. In IC design, the results are proprietary and most often kept secret. It is one of those fields where Kerckhoffs's law does not hold [70]. Obscurity is of utmost importance in the arms race between secure IC designers and reverse-engineering attackers.

3.6 Software Tamper-Resistance

3.6.1 Introduction

Securing software is a difficult, if not impossible, task. This is especially true in the case of DRM, where only a few elements can be trusted. When writing secure software, the developer must remember the following recommendations:

- secure the weakest link;
- practice defense in depth by combining many complementary protections;
- fail securely by steering the system in the right direction;
- fail subtly by avoiding brutal disruption that leak information;
- to follow the principle of least privilege;
- compartmentalize in order to isolate potential breaches;
- use layered protections to avoid single point of failure; and
- keep it simple although he may need to add complexity in some parts of the software for obfuscation [129].

All these recommendations are mandatory. This non-exhaustive list demonstrates all the complexity of the task of the security designer.

Usually, DRM clients may execute in hostile or at least non-trusted environments. Therefore, tamper-resistant software is one useful extension of secure software. Tamper-resistant software for DRM mainly has three objectives [130]:

- ensure that the DRM client protects itself against tampering by a malicious host, in other words the code integrity as well as execution integrity must be preserved;
- ensure that the DRM client can conceal the program it wants executed on the host, in other words code privacy must be preserved; and
- ensure that the DRM client can decrypt information without disclosing the corresponding decryption key, the secret keys must not be revealed to the host.

Tamper-resistant software is a science still in its infancy [131]. Currently, designing tamper-resistant software and secure software is more black art than true science.

Before continuing, a cautionary note: those three objectives cannot be plainly fulfilled. Tamper-proof software is a myth [132]. Tamper-resistant software, with less ambitious objective than tamper-proof software, is just a barrier to entry for attackers. The question is what the height of the fence will be, and what the cost to implement and to maintain such fence will be. Nevertheless, software tamper-resistance is a useful tool, as long as the designer keeps in mind its limits.

Typically, the attacker who wants to defeat software proceeds using a four-phase scenario [133]:

1. *Analysis*: In this first phase, the attacker tries to understand the behavior of the software. In some cases, this phase is sufficient, for instance, to steal intellectual

property (such as implemented algorithms protected by trade secrets) or to extract secret information. Analysis requires two steps. In the first step, the attacker uses all available means to collect information. In the second step, the attacker draws the workflow of the software using the collected information. These two steps are called reverse-engineering. The analysis may be static or dynamic. Static analysis means that the analysis of the software and data occurs while the software is not running, whereas dynamic analysis occurs while the software is executing. If the software is not properly protected, static analysis is easier and faster but less accurate than dynamic analysis. Nevertheless, static analysis is sometimes sufficient to defeat a protection.
2. *Tampering*: With the information collected in the analysis phase, the attacker attempts to modify the software so that it acts in the way required by the attacker rather than in the way originally expected by its designer. Tampering may be static or dynamic. Static tampering modifies the code in the executable file, for instance, changing an instruction set to bypass a test of validity of password, so that on all future executions, modified code is loaded and executed. Dynamic tampering modifies the code or data during the execution of the application, using effected through a dynamic attachment attack using a debugger or a shared library.
3. *Automation*: Once the attacker has successfully tampered with the application, he/she may develop an application that automatically reproduces the exploit. This is the case for tools such as `QTFairUse6`, `jHymn`, or `FairUseWM`, which broke Apple's FairPlay and Microsoft's Windows DRM V10 [134] (Chap. 7, Protection in Unicast/Multicast). In most of the cases, the automatic exploit tool requires no specific skills and can be used by anybody.
4. *Distribution*: The automated tool is disseminated through websites or bulletin boards, allowing many people to use it. This type of attack is called BOPA: Break Once Play Always. With the Internet, it is impossible to stop this illegal dissemination.

Some scientific communications related to tamper-resistant software assume that the host is trustful. Unfortunately, as already mentioned, this assumption is wrong in the video content protection context. It is commonly recognized that complete protection of the application under the assumption that the host is not trustful (malicious host) is impossible [135]. Nevertheless, new techniques allow reaching some degree of protection even in a hostile environment [136]. Therefore, the goals of the designer of tamper-resistant software are the following:

1. Minimize the collected data.
2. Make reverse-engineering difficult.
3. Detect or, better, prevent tampering.
4. If the software is tampered with, either repair it or act in a defined way such as stopping, degrading behavior, and alerting a remote site.
5. Prevent repeatable hacking; the scope of the hack should be limited at best. The worst case is a hack that affects all the deployed systems. In this case, the hack is called a class attack.

Clearly, tamper-resistant software does not comply with Kerckhoffs's law. Software tamper-resistance is based on security by obscurity. If the attacker is aware of the traps and pitfalls laid by the designer, then she may easily circumvent all of them. The remainder of this section provides a basic introduction to these techniques and to the corresponding mindset.

3.6.2 Prevent Analysis

The first tool used by the attacker is the debugger. The main purpose of a debugger is to help the software developer locate the bugs in his code. A debugger offers many useful features for dynamic analysis such as reading the internal registers of the Central Processing Unit (CPU), stack, and external memories, modifying registers, instruction-by-instruction processing, and setting breakpoints. A breakpoint is a position within the software. Once the software reaches the breakpoint, the debugger interrupts the execution of the code. Then the developer/attacker can analyze the environment at this break point and scour the software step-by-step from that point on. He can even modify data in memory. Debuggers are precious tools for software developers. Examples of debuggers are the Windows Kernel debugger SoftIce [137], which is not any more commercially supported, IdaPro [138], the shareware OllyDbg [139], and the open-source GNU Project Debugger GDB [140]. A new trend uses "undetectable" instrumentation tools running at ring 0 in a hypervisor [141]. This approach uses the benefits of virtualization. Obviously, debuggers are also powerful tools for attackers. Therefore, the first countermeasure against reverse engineering is to detect the presence of a debugger. There are many types of detection techniques:

- Detecting race conditions that a debugger would not be able to support [142]. The developer uses real-time conditions that can only be met when executing at nominal speed. Debuggers necessarily slow down execution time; hence, they may not meet the race conditions. This countermeasure does even not require that the software reaches a break point. Implementing race conditions is extremely tricky and requires deep knowledge of the behavior of the software.
- Detecting the presence of concurrent processes. One potential trick is to let the software try to debug itself. Were a debug process already running, this would fail [143].
- Detecting the use of debug-dedicated registers. Some registers of CPUs are dedicated to debugging. It may be interesting to detect their use or, better, to spoil their values, which would fool the debugger.
- Detecting the presence of backdoor communication channels [144].

Unfortunately, virtualization has brought its load of problems for the defender. Virtual machines can spy on the application without instrumenting the code. The first new challenge is to detect that the application is running on a virtual machine. Joanna Rutkowska's Blue Pill rootkit introduced the first proof of the concept of stealthy execution of a virtual machine [145]. The stealth of virtualization is one of the hottest debated problems in the field of reverse engineering. Today, the consensus is that a stealthy virtual machine is probably impossible. Once it detects that the application is operating on a virtual machine, the application should decide if it is legitimate. This is not a trivial answer and heavily depends on the context.

Once the protected application detects the presence of a debugger, the second countermeasure is to penalize it. The trivial approach is to terminate the program. Unfortunately, this termination is a clear indicator of the presence of an anti-debugger. More sophisticated responses may slightly change the behavior of the application, for instance, by introducing errors that would only occur later in the execution or by decreasing performance. In other words, they do not block the debugger but try to fool the attacker by acting in a non-generic way.

Some debuggers have been specifically designed for reverse engineering. For instance, RR0D runs beneath the operating system [146]. Thus, its presence is extremely difficult to detect.

Efficient detection of debuggers is intimately linked to the targeted debugger. Thus, a detection method may work for one debugger and not for another. The detection may even fail with a different version of the same debugger. Once more, it is an arms race between the reverse engineer's/debugger's designer and the secure software engineers.

The second tool used by the attacker is the disassembler. A disassembler translates from binary code into assembly language (whereas an assembler translates from a high-level language into binary code). Although it is at an extremely low level, i.e., with low semantics, assembly language can be understood by humans. Assembly language offers the highest level of control to the developer. Disassemblers format the binary code in a human-readable format. In some cases, if corresponding data are available, disassemblers can replace binary labels and data with the corresponding symbolic information.

A possible countermeasure is to try to fool disassemblers and decompilers. It mainly consists of adding junk code that does not affect the behavior of the application. When accessing this junk code, the disassembler will be lost. Nevertheless, disassemblers resynchronize after a while. Thus, it is mandatory to spread junk code all over the software. Several techniques are used, such as predicates [147] and overlapping instructions [148]. Encrypted software is also a good deterrent.

Of course, another technique is to decrease the readability of the software. For that, the developer should strip it of all meaningful information. All symbolic information has to be removed. In good software practice, the name of variables and functions should carry meaningful information, which help software designers maintain the code. This information is also useful to the hacker. Therefore, it is mandatory to strip away or encrypt all textual symbols, all chains of characters that indicate dependant products, and of course all debugging information. For instance, the function `void myFoo(int x, int y)` becomes `Type_0 F_1(Type_2 X_2, Type_3 X_3)`. This simple and cost effective technique offers several advantages:

- The attacker has far less insight into the handled data and the implemented features. It is obvious that a function labeled `0xFF31908D` is far less meaningful than the same function labeled `VerifyEncryptedPassword`.
- The disassembler cannot replace the addresses of functions, calls, and memory locations by their semantic meaning. The attacker will have to work with hexadecimal addresses.
- It does not affect the performance of the software, nor introduce bugs.

Another useful technique is code encryption. We will come back to this technique in the next section. Encryption serves two purposes. Confidentiality makes the analysis difficult. It is easier to understand a clear code than an encrypted one. Integrity thwarts tampering.

All these features blocking debuggers and disassemblers penalize also legitimate developers. If ever he detects a bug in the code, he will not be able to use these tools to debug his software. Sometimes, simply disabling the countermeasures may hide the bug. This may be the case for dangling pointers, i.e., pointers pointing to a wrong position.

3.6.3 Prevent Tampering

To prevent tampering, the first goal of the DRM designer is to prevent the attacker from reading the code after her successfully defeating the anti-debugger and the anti-disassembler. There are many methods.

The first technique is code obfuscation. Obfuscating program *P* means to generating a new program *P*' using a set of transformations (called obfuscators) O_i so that *P*' has the same behavior as *P* (i.e., *P* is semantically preserved). The obscurity of *P*' is maximized, i.e., reverse engineering *P*' is more time consuming than reverse engineering and understanding *P*. Furthermore, automatic tools should not be able to undo O_i, and the penalty in execution time and space should be minimal. An ideal obfuscation tool would turn a program code into a black box. Thus, it would be impossible to extract any property of the program. Unfortunately, such an ideal obfuscation tool does not exist.

In 1998, Christian Collberg, Clark Thomborson and Douglas Low [149] relaxed the constraints by stating that *P*' should have the same observable behavior as *P*, i.e., the user should perceive no differences. This allows *P*' to have additional side effects such as create new files, contact remote servers, or present a different error handling mechanism.

Many techniques of obfuscation are available. Among them, we list the following:

- Control transformations using opaque predicates. An opaque predicate is a section of code that generates a jump (conditional or unconditional). The predicates are always true, always false, or randomly false and true. For the developer, these predicates are deterministic because they are known at

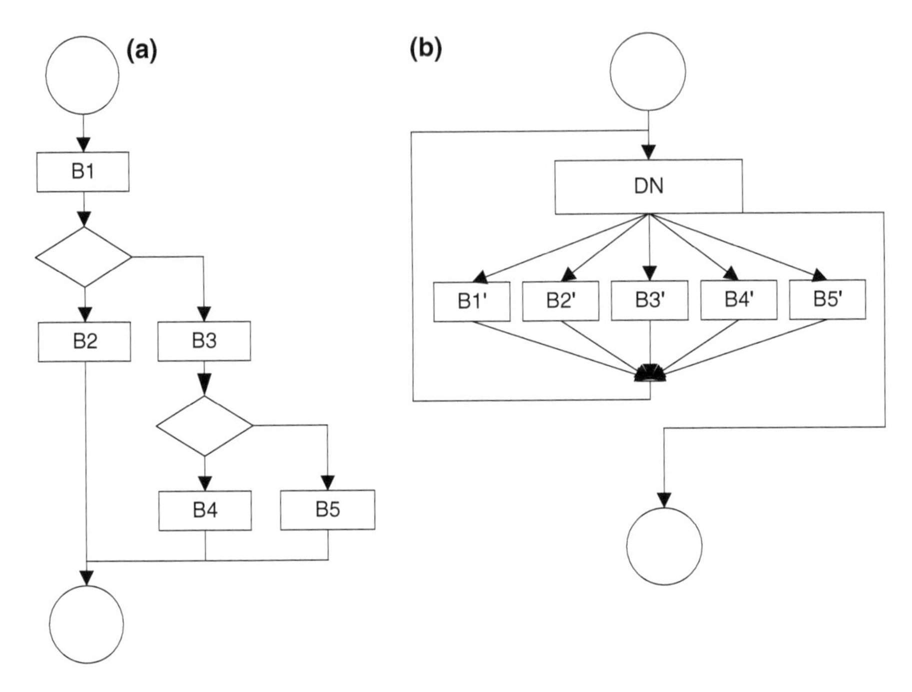

Fig. 3.6 Control flow graph flattening

obfuscation time. For the attacker, they are not deterministic and increase the complexity of the code.

- Insert irrelevant code using opaque predicates. This irrelevant code increases the complexity of the software to be reverse-engineered.
- Increase the test conditions of loops without changing the number of expected loops.
- Use concurrency to create opaque predicates; the technique uses different threads of execution to perform one function that could otherwise have been performed in one unique thread.
- Flatten the control flow graph; every program can be decomposed into a control flow graph of basic blocks with some test points. The technique of control flow graph flattening adds a dispatcher node that successively distributes the control to each basic block in order to simulate the sequencing of the unprotected program [150]. Figure 3.6a provides an example of a simple algorithm. It implements two successive cascading if-else structures with basic blocks B_i. Figure 3.6b describes the flattened implementation. The modified blocks B_i' add some control information that the dispatching node (DN in the figure) uses to decide to which next basic block it should pass the control.

This method is robust against static analysis. There are some ways to increase the difficulty of reverse engineering using cryptographic hash functions, complex switches, and non-local variables using opaque predicates calculated somewhere else in the program [151].

The Devil's in the Details

In a secure piece of software, as usual, the main assets to protect are the keys. Unfortunately, this is one of the toughest challenges. A typical error of implementation is to store the keys contiguously, for instance, all the bits of an AES-128 key stored in 16 consecutive bytes of memory. This has proven to be disastrous. Unfortunately, the size of software and storage space is exploding, which may give the false feeling to developers that the keys are safe because the brute-force exploration of every possible combination of 2 GB of RAM is too complex. This is especially true in the case of asymmetric cryptography such as RSA, which uses larger keys. Modular exponentiations require a lot of calculation, making the brute-force attack extremely time consuming, and thus impractical.

Nevertheless, several methods of analysis can drastically reduce the space of keys to explore [152]. Shamir has presented an interesting approach [153]. He made the hypothesis that an RSA key would be contiguously stored within the program. He roughly evaluated the entropy of every 64 consecutive bytes. Keys are by nature random; thus, they have high entropy. A software program is structured language, and thus by nature has lower entropy. By analyzing the entropy of the memory space, Shamir identified the few potential locations of random data.

Thus, it is extremely pertinent to hide this entropy singularity. Several methods are possible. The first one is to spread the key in many locations of the program code. This leverages the entropy. The second method is to encrypt the program code and decrypt it only at execution. This method increases the entropy because encrypted code has high entropy. Thus, entropy analysis cannot discriminate code from keys.

These methods only prevent static analysis of the code. They do not prevent dynamic analysis. Secure software implementations have to use the full collection of tamper protection techniques.

Lesson: Handling keys is the most crucial task in security. Unfortunately, it is also the most complex one. In the case of content protection, the task is even more difficult because the attacker controls the executing environment. Hiding the keys will be the ultimate goal of the DRM designer.

A second method is code encryption. There are two types of code encryption:

- Static encryption encrypts all the code. Before executing the application, a piece of bootstrap code decrypts the entire encrypted code. This method is more efficient against static analysis and far less efficient against run-time analysis. Once decrypted, the complete image of the application is available in the clear in the RAM, and the application is thus vulnerable. Nevertheless, the method has the advantage of only slightly impairing execution performance, and only at the start. Once the code is fully decrypted, there is no more time penalty.

- Dynamic encryption encrypts not the full code, but only a set of small pieces of code. These pieces of code are only decrypted when needed at run-time. This method is more efficient against dynamic analysis. At any time, only a small part of the code is in the clear in RAM. For instance, it is extremely difficult to put a software break point in the encrypted part. However, it is still possible to put hardware break points. The attacker has to stop the execution to access the clear code. Thus, a countermeasure against break points is ensuring that the application is continuously running. Unfortunately, dynamic encryption has a cost on performance. Execution time incurs up to a 100 times slowdown [154]. It should only be used to protect crucial functions.

The Devil's in the Details
Fault injection is a complex but devastating class of attack. The purpose of fault injection is to make the software act in a different way than expected through external stimulus in the environment. This attack is not intrusive in the sense that it does not require us to modify the software or use a debugger. The stimulus should generate a fault in the processor, resulting in a program execution that does not work properly. A typical example is forcing a conditional branching decision in a direction different from the logical one. The stimulus could, for instance, a quick drop of voltage in the power supply, "hiccups" in the clock signal, or a low level of ionizing radiation [155]. For example, replacing the transient edge of a 5 MHz clock of a smart card by one or several 20 MHz pulses may cause wrong instructions, due to the different gate delays and different paths in the processor.

The following example should help us understand the concept of fault injection and how it is possible to counter some attacks. We propose an oversimplified implementation of the verification of a four digit PIN (Personal Identity Number). The verification allows for three successive failed trials even if between two trials the card has been reset. The procedure returns Boolean information indicating if whether access is granted, i.e., the PIN is valid. The implementation uses two global variables stored in the non-volatile memory of a smart card: `storedPin` holds the exact value of PIN, `trialCounter` memorizes the numbers of successive failed trials. A flag, `blockPIN`, if set to "1", blocks definitively the smart card. The following is a possible implementation.

```
Boolean test_PIN(int dialedPINT);
Boolean test_Pin (int dialedPIN) {

    Boolean testFailed = TRUE;
    if (dialedPin == storedPIN) {

        trialCounter = 0;
        testFailed = FALSE;
```

```
        } else {
                if (trialCounter >=3)
                        blockPIN = TRUE;
                else
                        trialCounter ++;
        }
        return testFailed;
}
```

Figure 3.7 roughly illustrates the proposed implementation.
Where does this implementation fail? If the attacker identifies the time when the test on the trial counter occurs, she may succeed in injecting a fault into the test, i.e., she makes it fail. Then, the attacker may perform as many tests as possible because the trial counter will never be incremented. Figure 3.8 proposes a more secure implementation. The increment occurs before the test. Thus, there is no interest in attacking the test for counter, especially if its value is redundantly tested elsewhere in the software.
Lesson: Safe failing is mandatory in the implementation. Add also a salt of redundancy when checking critical conditions.

The second goal of the DRM designer is to prevent the attacker from modifying his/her code. There are only few mechanisms to achieve it. The main idea is that some parts of the code, called hereafter guards, check the integrity of some other part of the code. Of course, the weakest point is the guard's source code. Therefore, another guard verifies the integrity of the previous guard, and so on. At the end, a web of guards protects the software, where every guard checks the other guards. To be resistant, the guards must be diversified and difficult to detect in the source code. If guards would always use the same source code, then automatic pattern detection would locate them. Once detected, guards are rather easy to deactivate. An interlaced web of guards obliges the attacker in deactivating all of them. If well designed, the web of guards may make things hard for the attacker.

Commercial solutions are available. These tools range from fully automatic to semiautomatic, where the designer has to build his/her web of guards [156, 157].

Obviously, encryption is also a first fence against software tampering. Even if the attacker succeeds at decrypting the code, she cannot alter the encrypted code. She has to operate on the clear code, i.e., in real time. It means also that if she expects to distribute the patched software, she will have to distribute the hacked software in the clear, and the complete package rather than a cracking patch. In some cases, this may be a problem.

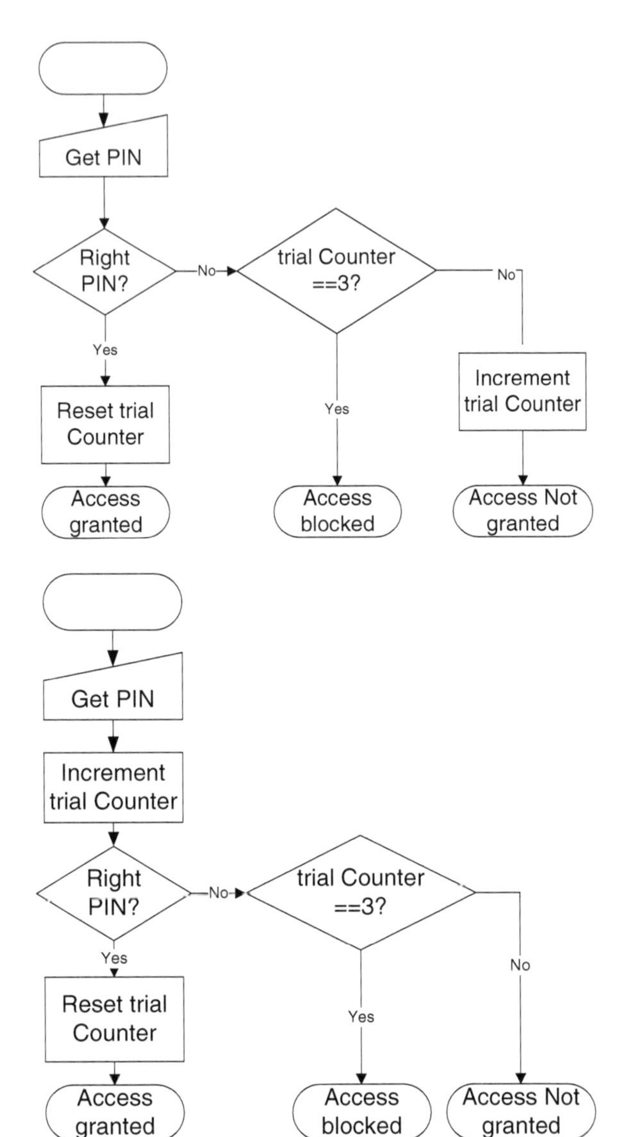

Fig. 3.7 First implementation of PIN test

Fig. 3.8 A better implementation

Tamper detection, which triggers corrective action, is also an interesting tool against attackers. The obvious reaction is to stop the execution once tampering detected. Unfortunately, this is the worst possible reaction for the defender. Stopping immediately the execution provides a direct clue to the attackers where the final test occurred. Thus, it is more interesting to use diversified reactions. For instance, it is possible to delay halting the execution. If, furthermore, this delay is random, it is even better. The triggering may also start a slow decay of performance of the software, or deactivate some features. In some cases, the action may even attempt to heal the software, or to modify the software to be different from the actual version.

We have focused on how to protect the code from tampering. Nevertheless, attackers sometimes succeed by modifying the data rather than the code itself. Protecting data is more complex than protecting code because data are by nature dynamic. This is a stimulating field of investigation.

3.6.4 Prevent Automation and Distribution

Unfortunately, once the attacker succeeds in breaking the application, i.e., building an instance of the application that acts as the attacker expects, it is extremely difficult to stop automation and distribution. Sometimes, the first issued version is rudimentary and does not have a nice, slick user interface. It may even operate only through command lines. Nevertheless, if the pirated software is interesting to many users, soon more user-friendly versions will appear.

The distribution channels of pirated software are the same as those of pirated content, i.e., ftp sites, P2P, and DDL sites.

There is potentially one solution limiting the spreading of automatic exploit tools. It is called diversification [158]. Diversification in software engineering means that each installed copy is unique. Alice's instance of the application is binarily different from Bob's instance of the same application, although both instances have the same functionality. Many techniques are available for diversification [159]:

- Replace one sequence of instructions by a functionally equivalent one.
- Change the order of occurrence of instructions; sometimes, the order of instructions can be changed without affecting the final result (for instance, with commutative operations).
- Modify the control flow by changing its complexity and adding some jumps.
- In-line functions or split them.
- Use different keys when encrypting the code.

The expectation is that an exploit working on one instance may not be reproducible on another instance. This has two consequences:

- The attacker can with difficulty issue a patch that would modify the protected application due to the diversity.
- Attackers cannot easily collaborate in the exploit because each instance of the application is different and thus presents a different configuration of protection.

Unfortunately, this type of protection has one serious drawback: the distribution of the genuine application. Each instance has to be individualized; thus, it is not suitable for mass distribution.

A second form of diversification is paramount and most probably mandatory: diversification between two versions or different products. The same protection should never use twice the same implementation. The same implementation defeated on one product/version will become useless if used in other versions. The attacker will know how to break it. Furthermore, attackers exchange a lot of information on specialized forums and conferences (such as Black Hat or DefCon) and the method that defeats the protection will soon become common knowledge in the community.

As a conclusion, software tamper-resistance is a new field of security, not yet mature. The protection is efficient against average users. Against skilled attackers, all protections will sooner or later collapse. It is only a matter of time and effort. Once cracked, the attacker can write a simple program to install the cracked version for average users [160]. The designers will have to build a new, more robust protection scheme.

3.7 Rights Expression Language

Rights Expression Languages (REL) describe the usage rights of a protected piece of content, i.e., what can be done with it. Thus, REL must have the following characteristics:

- It must be machine-readable because it has to be translated by the DRM agent.
- It must be expressive enough to support the requirements of the content owners, service providers, and merchants.
- It must be formal and unambiguous. It must be able to translate from commercial rights into usage rights enforced by DRM and vice versa.

Typically, commercial rights are expressed in natural languages, i.e., readable and intelligible to human beings. Commercial rights describe the contract between two parties. For example, commercial rights can be "a monthly subscription to "Syfy" channel" or "view the movie five times this year or make two copies." The same commercial rights may be expressed in natural language in many ways, whereas usage rights in REL must be expressed in one single and unambiguous way. In addition, it is essential to be able to translate usage rights expressed in REL into commercial rights expressed in natural language, for instance, to explain to the consumer what he/she may expect of the corresponding piece of content.

Typically, a REL expresses usage rights in a license by defining the following basic elements:

- The license describes the agreements between the party that grants the rights and the party that receives these granted rights.
- The issuer is the party that grants the rights. It may be the rights owner of the piece of content or an entity that acts on behalf of this rights owner.
- The subject is the party to whom the rights are granted. There must be ways to authenticate the subject, i.e., to prove that the subject is who he/she purports to be. Enforcement of REL mandates authentication. Nevertheless, the definition of the authentication method is outside the scope of REL.

- The resource is the "object" to which the rights are granted. The resource may be a piece of content, a service, a set of data, or even a physical object.
- A right describes the type of access to the resource that a subject can be granted. The right expresses actions like "print", "view", "edit", "copy", or "transfer". They can be classified into four categories [161]:
  - Render rights define representation actions such as print, view, or play.
  - Transport rights define the ways a piece of content may be space-shifted, such as copy, move or loan
  - Derivative work rights define the ways a piece of content may be used to create another piece of content, such as extract, embed, or edit.
  - Utility rights define rights that are more driven by technological rationale than business rationale, such as back up copy or temporary buffer.
- A right may have attached conditions and obligations called attributes. They define additional conditions, for instance, on the environment. Conditions limit the way the actions of the rights can be performed, for instance, "view three times within 48 h". Obligations define some preliminary conditions that need to be true before exercising the granted rights (of course, if the conditions are fulfilled), "allow output in High Definition (HD) if output is protected by High-bandwidth Digital Content Protection (HDCP), else output in Standard Definition (SD)".

Most probably, Mark Stefik, from XEROX's Palo Alto Research Center (the mythic XEROX PARC), was the founder of modern RELs in the 1990s [162]. The outcome of his early works is the Digital Property Rights Language (DPRL). It defines fees, rights, and conditions for accessing digital works. The first version was issued in March 1996 and was only machine-readable. The second version was released in May 1998. In addition to including new conditions and rights, its main evolution was that this language became declarative. It used the markup language concepts defined by XML DTD [163]. It became also human-readable.

In 2000, XEROX PARC spun off the company ContentGuard[16] to develop DPRL. The result is the eXtensible rights Mark-up Language (XrML) [164]. The new name reflected the adoption of XML to express the language. The first version was published in May 2000. In November 2001, the second version was released [165]. It became the core architecture and base technology for MPEG standards MPEG-21 Part 4 Intellectual Property Management and Protection (IPMP), MPEG-21 Part 5 REL, and MPEG-21 Part 6 Rights Data Dictionary (RDD) [166]. Despite the fact that MPEG-21 is an ISO standard, eight years after its creation, MPEG-21 has not yet been widely used in commercial applications.

With XrML vocabulary, a content owner provides a grant. The grant describes four elements:

[16] ContentGuard is co-owned by Xerox, Microsoft, Time Warner, and Technicolor.

```
<xsd:schema   targetNamespace=http://www.example.org/interact
xmlns:i=http://www.example.org/interact
xmlns:r=http://www.xrml.org/schema/2001/11/xrml2core
xmlns:dsig=http://www.w3.org/2000/09/xmldsig#
xmlns:xsd=http://www.w3.org/2001/XMLSchema elementFormDe-
fault="qualified" attributeFormDefault="unqualified">
        <xsd:import names-
pace=http://www.xrml.org/schema/2001/11/xrml2core />
        <xsd:element name="accept" type="i:Accept" substitution-
Group="r:condition"></xsd:element>
   <xsd:element name="register" type="i:Register" substitution-
Group="r:condition"></xsd:element>
   <xsd:complexType name="Accept">
   <xsd:complexContent>
        <xsd:extension base="r:Condition">
                <xsd:sequence>
                        <xsd:element name="statement"
type="r:LinguisticString" MaxOccurs="unbounded"/>
   </xsd:sequence>
                </xsd:extension>
   </xsd:complexContent>
   </xsd:complexType>
        <xsd:complexType name="Register">
   <xsd:complexContent>
                <xsd:extension base="r:Condition">
                        <xsd:sequence>
        <xsd:element                name="registerServiceReference"
type="r:ServiceReference"/>
                        </xsd:sequence>
                </xsd:extension>
   </xsd:complexContent>
   </xsd:complexType>
   </xsd:schema>
```

Fig. 3.9 Example of an XrML license

- The principal is the entity that receives the grant. The principal is uniquely defined and needs to be authenticated.

- The right is an action or a set of actions that the principal is authorized to perform.
- The resource is the object on which the principal may apply the granted rights. The resource may be a digital object, a service, or a piece of information such as an email address.
- Conditions define the set of conditions to be met to exercise the right. Conditions may be, for instance, a time interval or a given environment of execution.

The grant is defined within a license. The license is generated by the issuer who digitally signs the license to prove the origin of the license. Figure 3.9 presents an example of an XrML expression.

The main contender of XrML is the Open Digital Rights Language (ODRL). The current version 1.1 was published in August 2002 [167]. Since 2008, a new version ODRL 2.0 has been under development [168]. The Open Mobile Alliance (OMA) (see Sect. 7.5) uses ODRL as its REL. ODRL uses three core elements:

- The asset is the element on which the grant/permission is delivered. An asset represents physical or digital content. It may have many subparts. Each asset is uniquely identified.
- The party is an entity that either provides or receives the rights. Parties can be end users, roles, or rights owners. It is equivalent to XrML's resource.
- The rights define the permissions granted to an asset. Permissions define the following:
  - Constraints, which define limitations on the assets.
  - Conditions, which are limitations on the permission itself (such as cannot be resold)
  - Requirements which are preconditions that the environment or the party should respect prior to getting the permission.

In addition, ODRL defines the concepts of offers and agreements that bind the three previous elements (Fig. 3.10).

XrML and ODRL are declarative languages. Thus, they lack dynamicity. Furthermore, some opponents have complained that declarative languages are not able to describe complex copyright uses. A new breed of language based on logic and programming rather than declaration is appearing. The first instance of this new generation is LicenseScript [170], which appeared in 2003. LicenseScript claims it can approximate the "fair use" law [171]. There is currently no commercial deployment of these new logic-based languages.

3.8 Compliance and Robustness Rules

For a device to implement a content protection system, it has not only to implement technical specifications, but also to ensure that it correctly interoperates with other devices, servers, a list of supported formats that it understands and also a list

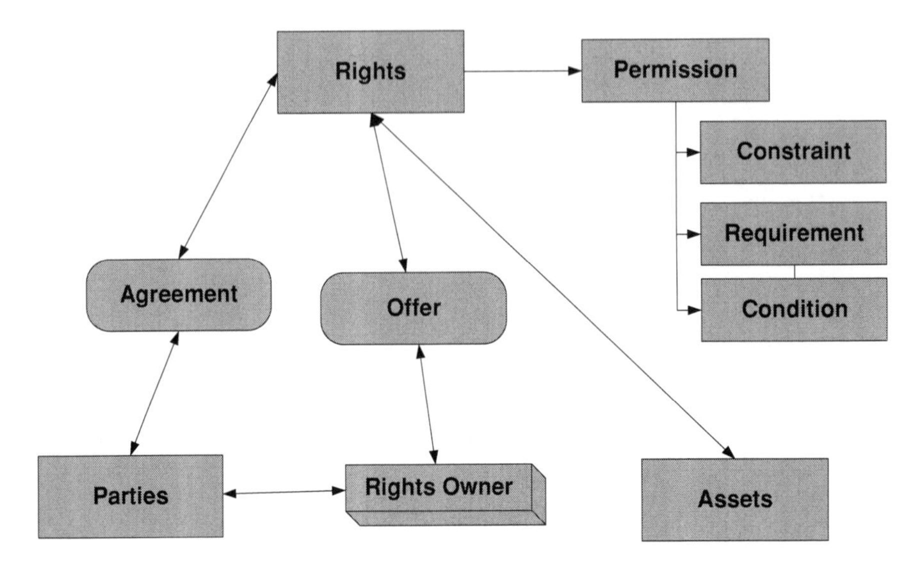

Fig. 3.10 ODRL information model [169]

of other content protection schemes that it can import from or export to. This is called conformance. The device also has to provide the adequate level of protection. Security means nothing if the device does not reliably and predictably enforce, enable, and guarantee behavior in accordance with expressed and implied rules and policies [172]. This is achieved by compliance with the robustness rules. The complete framework forms the compliance and robustness regime (C&R) [173]. Each video content protection scheme has its own C&R.

Security is the result of successive layers of constraints (Fig. 3.11).

Trust is the cornerstone of content protection. Trust means that only compliant principals participate in the ecosystem. Thus, it is paramount to define what a compliant principal is and how it should behave. This is the role of the compliance and robustness rules.

Compliance rules ensure that the principal will act as expected. They fully define the expected behavior. They often rely on technical specifications, but also on behavioral conditions that are part of the technical specifications. For instance, compliance rules may require us to follow usage rules contained in copy control information associated with the content, to use approved protected outputs, and to use approved protection recording technologies in those instances where consumer copying is permitted [174]

The following simple example will illustrate the importance of compliance rules in a scheme of content protection. In days of analog technology, video/audio tape recorders had a record button that enabled copying a piece of content. This record button has disappeared from DVD players. This is because the licensing agreement signed by DVD manufacturers forbids the implementation of such a

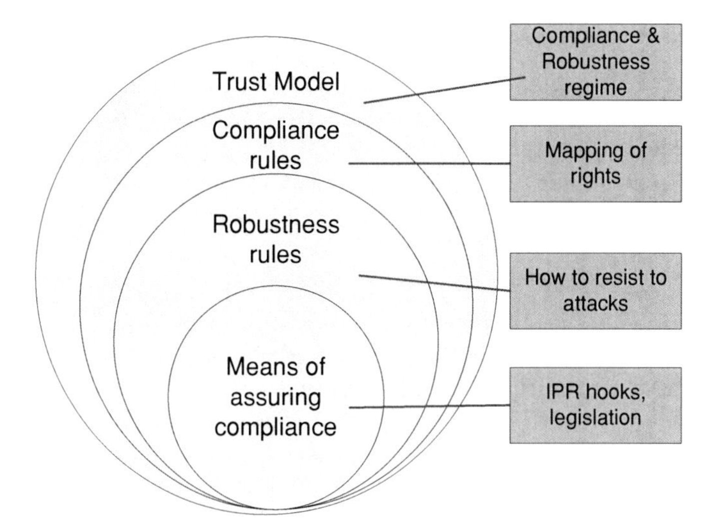

Fig. 3.11 Compliance and robustness regime

function. No technical barrier would have prevented designing a DVD player that could also act as a recorder.

Of course, specifications of compliance rules are more granular. For instance, for a piece of content marked as copy never, the compliance rule will express that such content should never be stored in persistent memory. Persistent memory encompasses optical storage, hard drives, and flash memories. The following is an example of such a compliance rule.

2.1. **Copy Never**. Licensed Products shall be constructed such that Copy Never DT Data received via their Sink Functions may not, once decrypted, be stored except as a Transitory Image or as otherwise permitted in Sect. 2.1.1.
2.1.1 Copy Never DT Data may be retained (i.e., stored) for such period as is specified by the Retention State Field, solely for purposes of enabling the delayed display of such DT Data. Such retained DT Data shall be stored using a method set forth in Sect. 2.2 and shall be obliterated or otherwise rendered unusable upon expiration of such period.

Within digital devices, video is often buffered, i.e., temporarily stored, to perform digital enhancements. This usage is legitimate, even for copy never marked content. Thus, compliance rules cannot simply ban storage. Thus, the above rules introduce the concept of transitory image. The compliance rules define the maximal period of retention to prevent misuse. In this case, a dedicated Retention State Field defines it. Often, lawyers and engineers write compliance rules hand to hand.

Compliance mainly defines three types of conditions:

- In what form a piece of content may be accepted by the device; this means both the format and the authorized types of copy protection schemes used to protect the incoming piece of content.
- In what form the device may export a piece of content; this means both the format and the authorized types of copy protection schemes used to protect the outgoing piece of content.
- The authorized actions for the piece of content.

Compliance rules define the exhaustive list of the copy protection schemes as authorized input and the list of the copy protection schemes as authorized output. Furthermore, compliance rules define the mapping of usage rights between two heterogeneous copy protection systems (Sect. 11.4.2, Usage Rights Translation). This is the cornerstone of current content protection. Today, there is not one single technology that protects the work from end to end. It is rather protected by a set of successive content protection schemes acting in a trusted chain (or a trusted web). For instance, within the home network, many protection schemes will have to cooperate to protect a movie. A DRM may protect the movie delivered to a residential gateway. Inside the gateway, the corresponding resident DRM client will unprotect the movie and pass it to Digital Transmission Content Protection (DTCP) (Sect. 9.3). DTCP will protect the content during the transmission over Ethernet to the Digital Video Recorder (DVR). The DVR will use its Copy Protection for Recordable Media (CPRM) (Sect. 8.4) to protect the movie while it is burnt onto a disc. Once the DVR plays the movie back, it uses HDCP (Sect. 9.4) to protect the content over the High Definition Multimedia Interface (HDMI) connection. For this chain of content protection systems to work reliably, each content protection scheme has to trust the other ones. Thus, they create a kind of trusted ecosystem. Figure 3.12 illustrates such a chain of trust.

Robustness rules ensure that the device will resist to any attempt by crackers to alter its behavior or to extract secret information. For that purpose, robustness rules list the assets to protect. For each asset, robustness rules specify the level of expected protection against a set of potential attacks.

Robustness rules cannot literally prevent a device from being hacked. No security system is perfect. Nevertheless, robustness rules do make it more difficult. By conforming to such rules, a manufacturer demonstrates due diligence in protecting the assets. This is typically a licensing requirement for becoming a part of the ecosystem. The robustness rules are often rather generic. The interpretation relies on the manufacturers' goodwill. Nevertheless, the licensing authority always has the option to evaluate compliance with these rules before providing certification. Typically, robustness rules define three categories of tools:

- Widely available tools are very basic tools that are not expensive and do not require any specific skill. Examples of widely available tools are screwdrivers and simple file editors.

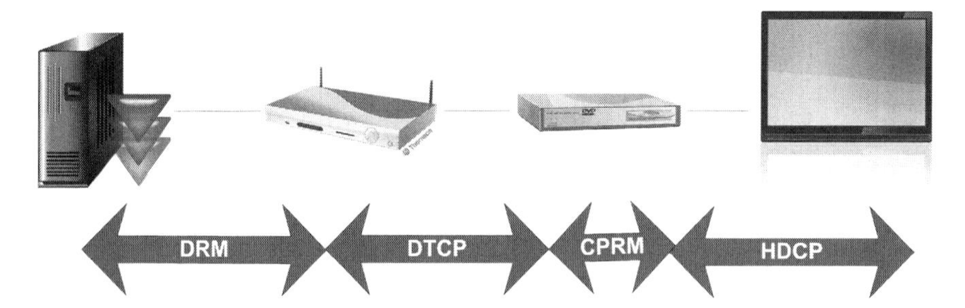

Fig. 3.12 Chain of content protection systems

- Specialized tools require more skills and are affordable (if not freely available on the Internet, as in the case of some software tools). Examples of specialized tools are debuggers, memory scanners, and memory dumpers. The second-hand market gives access to tools that were initially only available to specialized laboratories.
- Professional tools are expensive tools that require specific skills and training. Examples of professional tools are expensive disassemblers such as IdaPro, logic analyzers, and JTAG readers.

This taxonomy roughly maps to IBM's classification of attackers [120]. This classification is extremely interesting and can be widely used. IBM defines three types of attackers.

- Class I: Clever outsiders who are often very intelligent people. They do not have a deep knowledge of the attacked system. They have no funds and thus may use widely available tools. They will often exploit known weaknesses and known attacks rather than create new ones.
- Class II: Knowledgeable insiders who have specialized education. They have good, although potentially incomplete, knowledge of the targeted systems. They are likely to use specialized tools, and have some financial resources.
- Class III: Funded organizations that are able to recruit teams of complementary world-class experts. They have access to professional, expensive tools. They may even design their own custom tools. Moreover, they have "unlimited" financial resources.

Once the tools and attackers are classified, it is possible for each asset to define the expected level of security. For instance, Microsoft defines three types of asset [175]:

- Level 1 assets, such as device keys, have to be immune to widely available tools and specialized tools.
- Level 2 assets, such as content keys, have to be immune to widely available tools and resistant to specialized tools.
- Level 3 assets, such as profile data, have to be resistant to widely available tools and specialized tools.

Note the nuance between immune and resistant. Immune means that the attack should be impossible (at least with this type of tool), whereas resistant means that a successful attack should require much effort.

It is crucial to enforce that device manufacturers comply with the compliance and robustness regime. Only legal methods can ensure this. Most of manufacturers use patents as an enforcement tool. In addition to enforcing the obligation to pay royalties for the use of patented solutions, they enforce two other conditions:

- If a circumventing method is known in advance, then it may be interesting to patent, if possible, the method. Were a manufacturer to design a circumventing device using this technique, it would be possible to sue it for patent infringement [176]. In many countries, proving patent infringement is easier and faster than proving copyright infringement. This is true in the US, where patent infringement triggers immediate seizing of the infringing devices, whereas DMCA infringement cases are slower in court and do not allow immediate seizing.
- It is also interesting to add a patented solution that the system needs in order to work. This trick is called a hook IP. It provides the foundation for an antipiracy licensing strategy. The hook IP is only accessible when signing the licensing conditions. Thus, any unlicensed manufacturer that would build a device would necessarily infringe the hook IP. The signature on the product license implies the acceptance of compliance and robustness rules.

Thus, it may be difficult for a service provider to pursue a pirate for infringement of copyrighted content. Infringement of a patent as a hook IP may be easier to demonstrate than violation of DMCA.

For instance, if the Anti-counterfeiting Trade Agreement (ACTA) were to be agreed upon as the leaked draft indicates [177], there would be the obligation to seize "pirated" code at borders or in transit and to pursue users. Hook IPs would perfectly benefit from such an act.

In the past, Hook IPR has been successfully used to put legal pressure on "rogue" device manufacturers, confiscate unlicensed devices, and initiate legal proceedings. An emblematic example is the case by studios against the company Studio 321. Studio 321 designed software that allowed consumers to make backups of CSS-protected DVDs. Obviously, for that purpose, it had to break the CSS keys. Studios filed a DMCA lawsuit that took more than one year to win. Meanwhile, the CSS licensor and Macrovision filed a case for patent infringement because Studio 321 violated some of their patents. This action prevented Studio 321 from further selling its software [178]. Later, the lawsuit definitively banned Studio 321 from selling this product.

Chapter 4
Modeling Content Protection

4.1 Introduction to Different Models

To model a system, the designer often analyses different points of view [179]. Each model will represent one particular point of view. Using one single model is often insufficient for understanding the system. This is especially true for complex systems such as DRM. Currently, content protection models primarily address functional, transactional, and architectural points of view. The functional model describes the different functions of a system and the interactions between them. The transactional model defines the chaining of exchanged messages, of their steps. The architectural model maps the different elements of a system to actual principals such as servers, storage units, and devices.

4.2 Functional Model

The functional architecture of DRM has three main areas (Fig. 4.1):

- management of the rights,
- management of the content, and
- management of the usage

The management of rights has two functions [180]:

- *Creation*: A creator issues an original piece of content. This process is not only artistic, but also highly technical. For instance, the creation of movies implies many postproduction processes such as video editing, color management, audio management, special visual effects editing. The creator uses a content provider to distribute his work. In some cases, the creator can also be the content provider. The content provider defines the usage rights of a piece of content. These usage rights may vary for each consumer or may depend on a commercial bundle. Usage rights are not necessarily associated with a commercial offer.

E. Diehl, *Securing Digital Video*, DOI: 10.1007/978-3-642-17345-5_4,

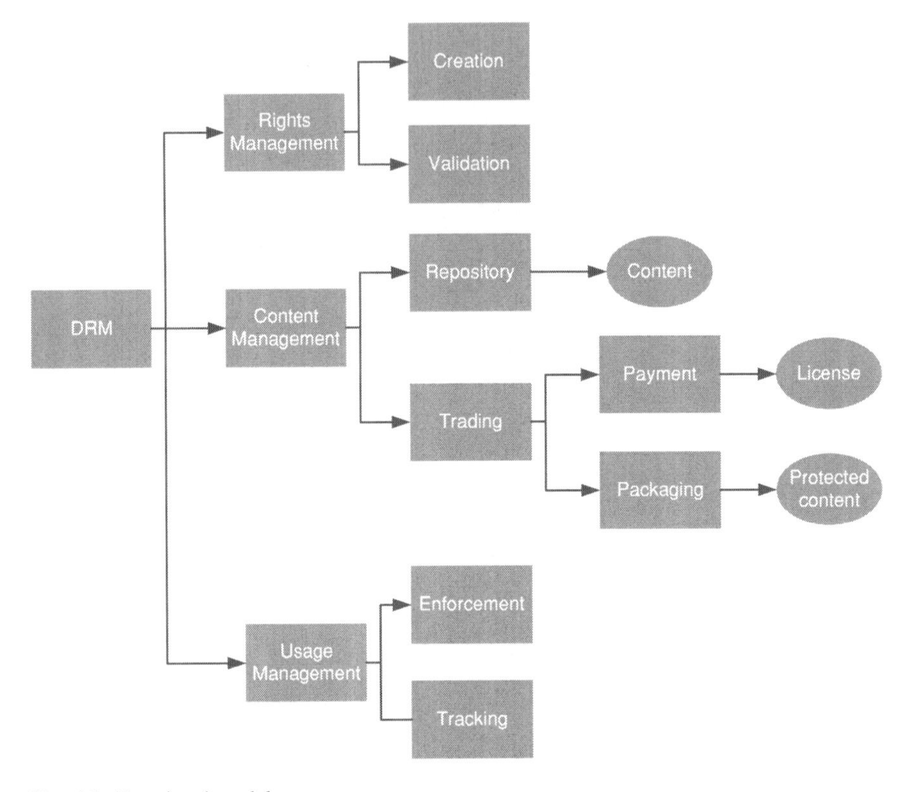

Fig. 4.1 Functional model

In the enterprise environment, DRM may protect documents from economic intelligence. Thus, it will tightly control the usage rights to prevent leakage. For instance, it defines who may access the document, who may modify it, and who has the right to send it.

- *Validation*: When a consumer attempts to use digital content, DRM checks the validity of her usage rights. It verifies whether the consumer has the credentials to perform the action she requests. If the consumer does not have the credentials, then the DRM will not allow the requested action. Optionally, it may provide hints on how to obtain the needed credentials.

The management of the content handles the essence of the piece of content itself. It has two main features:

- *Repository*: It manages a repository of all the available content that the DRM will protect. The repository may be a remote server or fully decentralized (for instance, using Peer To Peer-based contribution network).[1] The pieces of content are received from the content owners. Usually, contents are stored in an encrypted form to prevent theft or leakage.

- *Trading* provides the means to sell and deliver the piece of content to a consumer. This trading feature has two main tasks:
  - Packaging prepares the content for distribution. It mainly transforms the piece of content into a format supported by the consumer's environment and protects it by cryptographically binding it with the license. The transformation may change the resolution of the content, change the codec, and use a given container. A codec is an audio or video compression algorithm such as MPEG2, H264, or JPEG2000, as `.avi`,`.mov`,`.mp4`, or `.vob` [181]. In this book, we will use the digital asset management convention [182] of calling essence the representation of the content itself, as opposed to ancillary information such as metadata, licenses, or headers.
  - Payment ensures that the content owner will be rewarded for the usage of its content. Payment is a complex feature that encompasses the following steps:
    - Collecting money from the consumer; this may be done in many ways such as through account charges, credit cards, regular subscriptions, prepaid cards, and vouchers.
    - Generating a license for the delivered content; the license contains information that allows the generation of a message of the form, "This license issued by A allows user B to perform action C on resource D under conditions E," for instance, "This license issued by merchant A allows Alice to view movie F on her PC G during the next two weeks" or "This license issued by merchant B allows Bob to burn the song D three times". "Action C on resource D under conditions E" is the natural language form of the usage rights. The license uses a REL to express these usage rights (Sect. 3.7). The license also carries the information needed to unprotect the protected content.
    - Managing a clearinghouse to redistribute the collected money to the rights holders. The clearinghouse logs all transactions and redeems the corresponding benefactors.
    - Obviously, for DRM that handles payment-free content, this feature is reduced to the generation of licenses.

Usage management has two main features:

- *Enforcement*: Usage enforcement ensures that the consumer will honor the granted usage rights.

[1] In fact, the storage and distribution of electronic files with guaranteed quality of service is a complex topic. In the multimedia world, it uses often a complex architecture, called Contribution Distribution Network, based on intermediate caching servers that attempt to bring the files as near as possible to the customers.

- *Tracking*: Usage tracking monitors when needed the consumption of a piece of content. It may be useful, for instance, if the payment is defined by conditions of usage such as pay per time. Tracking allows us to share fairly the revenues or to audit business models. Tracking may be mandatory in an enterprise environment to audit who edited or viewed a piece of content and when it occurred.

4.3 Transactional Model

4.3.1 General Model

DRM handles mainly eight transaction steps (Fig. 4.2) [183].

1. Provisioning content; the content provider delivers the original content to the distributor. It may be digitally or physically distributed. Physical distribution involves removable hard disks, DVDs, or digital tapes.[2] The current trend is towards tapeless processes, i.e., only using electronic file exchanges.
2. Safely keeping the piece of content on the content server. It is often mandatory to protect the content against theft or alteration. Therefore, the distributor has to encrypt content as soon as it is provisioned.
3. Placing the offer; the piece of content will be offered on certain conditions and terms. Each piece of content may be sold with its own conditions and terms. The conditions and terms for the same piece of content may differ from one consumer to another. It should be noted that the offer is not necessarily commercial, i.e., linked to a monetized transaction. This is the case, for instance, for enterprise DRM.
4. Booking, which provides the means for the consumer to access/purchase the desired content. The consumer may accept or refuse the proposed offer. This is often done through a merchant portal.
5. Payment: Once the consumer has accepted the offer, he/she might need to pay. There are many possible payment options such as credit cards, account charges (for instance, PayPal, and phone operator bill), prepaid cards, and vouchers. This step does not exist for payment-free type.
6. Content preparation: Once payment succeeded, the distributor prepares the content for actual distribution. It may adapt the format or resolution to the consumer's viewing environment. Furthermore, it protects the essence and generates the license.
7. Content distribution, which delivers the content and its license to the consumer. Often the delivery channel does not need to be secure because the content and the license are self-protected. The delivery of content and license may not be simultaneous. Furthermore, they may use different carrying methods.

[2] In some cases, this delivery may itself be controlled by DRM [184].

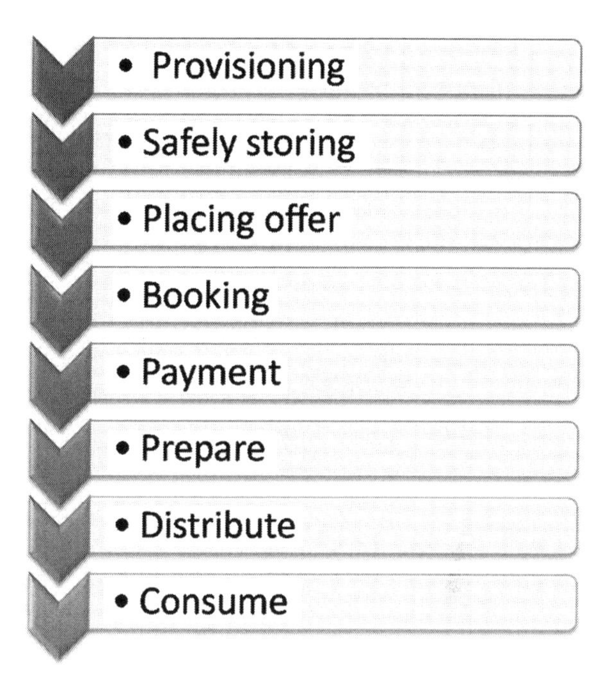

Fig. 4.2 Transaction steps

8. Content consumption, according to the conditions and terms defined in the license.
9. Redeeming, which shares the revenues among the different actors.

Steps 1–3 occur long before consumption. Steps 4–8 occur during actual consumption. Step 9 occurs after consumption. The order of these operations may vary. Sometimes, the protected piece of content is distributed before actual payment. In that case, only the delivery of the license occurs after the payment.

4.3.2 Payment-Based Distribution

Creation implies many actors, but from the DRM standpoint, we reduce it to one single role: content provider. The content provider delivers the original content to a distributor.

The distributor "sells" the content to consumers. The distributor informs the financial clearinghouse of the commercial offer associated with each proposed piece of content. The distributor generates the protected content and corresponding license before releasing it with its corresponding license.

The consumer needs both the protected content and its corresponding license to access the clear content. Thus, the consumer acquires the right to access the content from the access control manager, i.e., the merchant. The financial clearinghouse receives payment from the consumer and checks the validity of this payment. Once the payment is validated, the financial clearinghouse informs the access control manager of the successful completion of the transaction. The access control manager informs the distributor of the acquired rights. The distributor builds the corresponding license and delivers this license to the consumer. The license and the protected content can be delivered together (combined delivery), or they may be delivered separately (separate delivery).

A special case of separate delivery is superdistribution. In the superdistribution model, protected content is delivered without a prior content request. Consumers may freely disseminate the piece of content, and sometimes distributors may push it on the hard drives of consumers. Once consumer receives the super distributed content, he or she will place the content request with the identity of the already received content. The distributor only needs to deliver the license to the consumer. Superdistribution has the advantage of removing the burden/cost of the distribution from the distributor. Coupled with the social phenomena of viral distribution, many analysts believe that it may be the future preferred distribution mode.

One strong assumption is that protected content cannot be used without its corresponding license. Therefore, a distributor delivers the license only once the consumer has paid. Once the consumer has both the protected content and its license, he/she can consume content under the conditions defined by the acquired rights (Fig. 4.3).

Often the question is raised: "Does combined delivery provide any advantage to an attacker compared to separate delivery?" In other words, is protected content more secure when the attacker does not have access to the decryption key even while it is protected in the license? A similar question occurs with direct downloading versus file downloading. In other words, is file streaming more secure than file downloading? The usual argument with separate delivery is that the attacker may have earlier access to protected content, even if she does not yet have access to a legitimate license. Thus, the intuitive answer is yes. Unfortunately, intuition is rarely the best judge for security evaluation. If the trust model assumes that bulk encryption, i.e., the scrambling of the essence, is secure, then the right answer is no. Every trust model of content protection assumes that bulk encryption is secure. Therefore, a proper system design should assume that the attacker has full access to the protected content. Thus, there should be no difference from the security standpoint between a downloaded file and a streamed file, and between separate delivery and combined delivery. Assuming that streamed content is more secure than downloaded content is incorrect.

Nevertheless, in some cases, it may be wise to separate delivery of protected content and license. Let us suppose that the attacker finds a way to extract from one license the keys protecting the content and that these keys are common for every user. If the protected content is distributed independently from the license, then the attacker needs just to distribute the keys, i.e., she needs a smaller bandwidth for delivering the pirated material than for distributing the full content in the clear. In that case, once the consumer gets the pirated keys, it will be up to him to

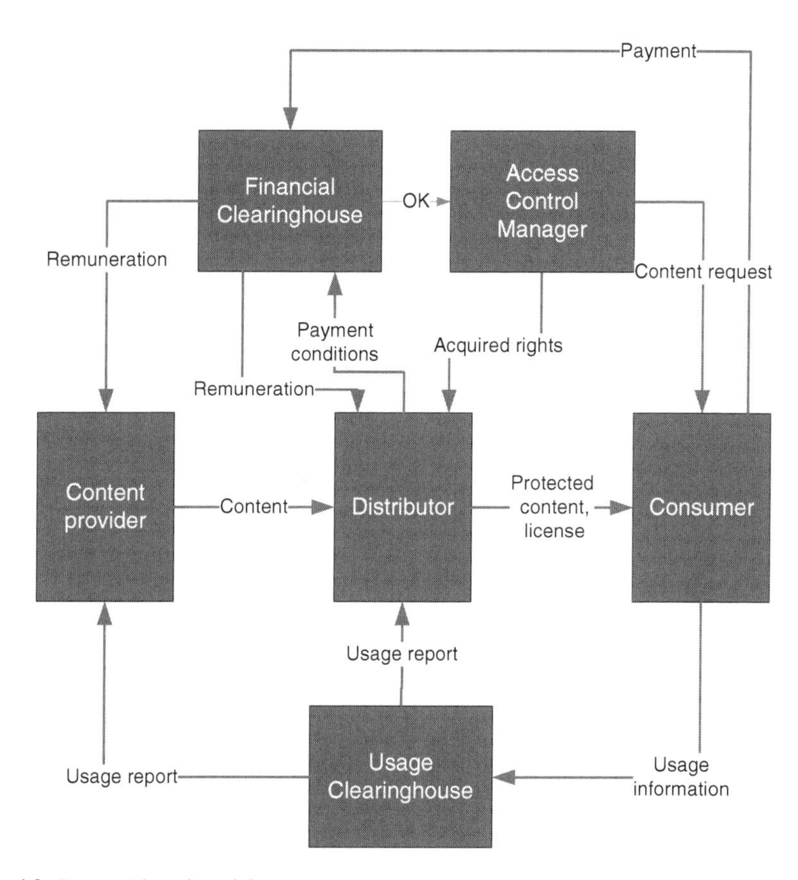

Fig. 4.3 Payment-based model

find the full-protected content. This was the case for the first hacks on HD DVDs [185]. The attacks just distributed the keys protecting the discs. With traditional attacks, the attacker has to distribute the complete content in the clear, thus requiring more bandwidth (see "The Devil's in the Details" in Sect. 11.2.2).

The last step is the payment to the content provider. The financial clearinghouse collects payments from consumers and distributes the revenues among content providers and distributors following rules that were contractually established.

A usage clearinghouse collects information from the consumers, analyses this information and dispatches it to content providers and content distributors through usage reports. Usage reports may serve several purposes, for example to better understand the consumption habits of consumers to fine-tune the commercial proposal and to validate the fairness of payments. Gathering data from consumers may violate privacy norms and consumers' expectations. Thus, special caution is needed for this collecting phase [186]. In 1999, RealNetworks distributed its media player software, which stealthily collected information about consumers' musical preferences [187]. In addition to the songs listened, the software returned the

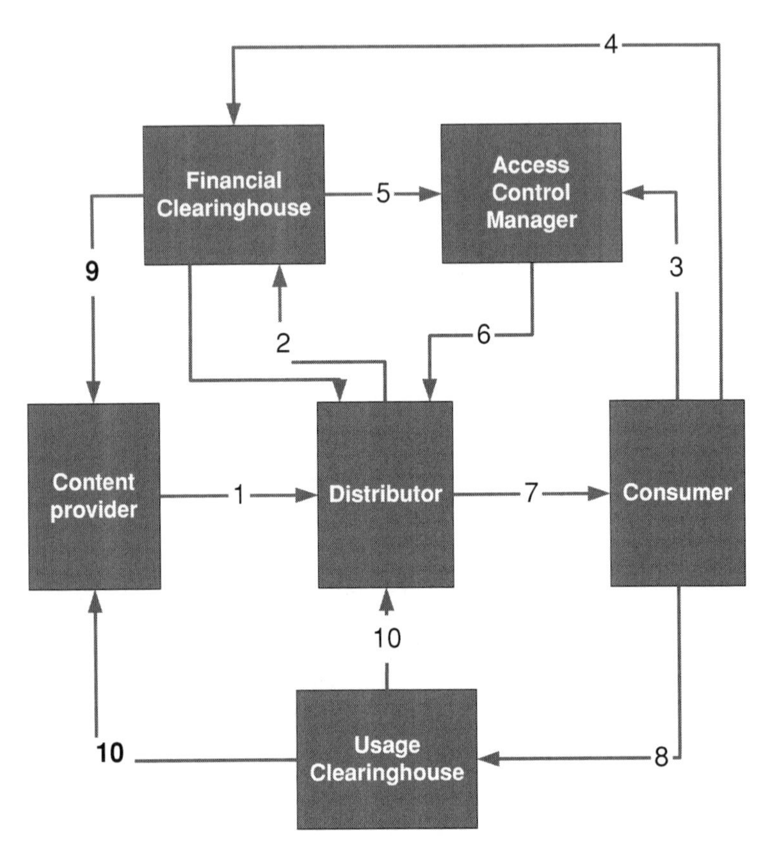

Fig. 4.4 Dynamic transitions

Generic User ID (GUID) of the hosting computer. This was done without the consent or the knowledge of the customers. Once this practice was revealed by the media, consumers reacted indignantly. Quickly, RealNetworks released a patch that disabled this feature. In the future, respect of privacy will become one of the most important expectations from consumers.

Figure 4.4 summarizes the scheduling of the transactions. There are mainly three phases.

The preparation phase (1–2): 1 corresponds to the provisioning and securely storing step. Step 2 corresponds to the placing offer step. There are no real-time constraints.[3] Delivery of content and definition of the commercial offer are done before the actual first distribution. Thus, these steps are often done off-line.

The transaction phase (3–7): This phase starts when the consumer requests a content (3), i.e., booking step. The operations must run quickly to offer a

[3] This is not true if DRM protects live content. In that case, the definition of the sales conditions occurs before the actual delivery of content from the content provider. Of course, the preparation of the protected content then has to be performed in real time.

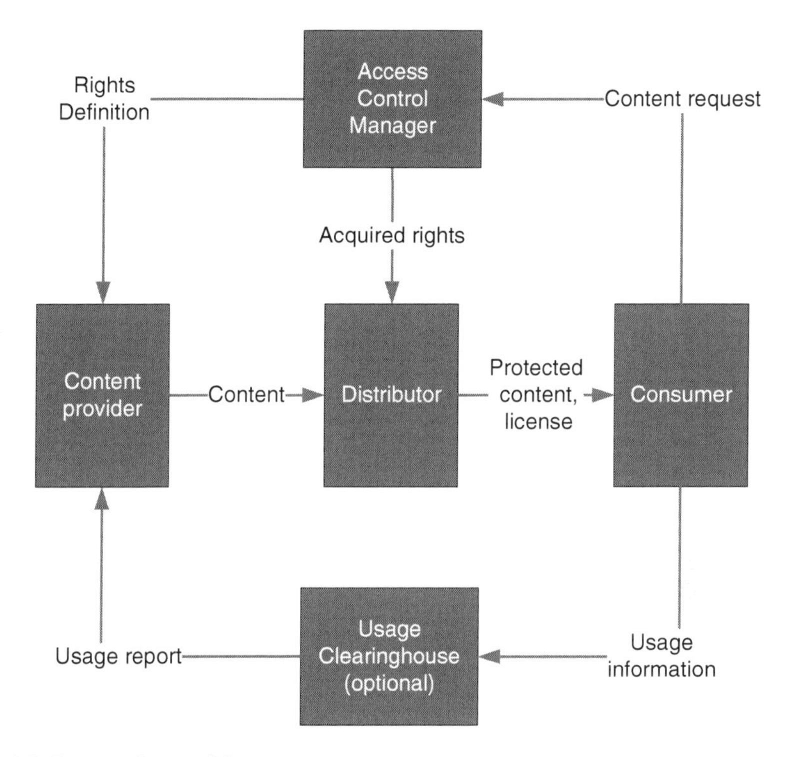

Fig. 4.5 Payment-free model

satisfactory experience to the consumer, especially once she has paid for her content (4). Steps 5 and 6 initiate the preparation step. The actual content delivery (7) has to occur as soon as possible.

The redeeming phase (8–10); the usage clearinghouse collects usage information (8). The collection may occur in real time each time the consumer accesses one piece of content or it may be collected on a regular basis. In this case, the consumer's client software logs the consumption until the next report. Remunerations (9) and usage reports (10) are done on a regular basis.

4.3.3 Payment-Free Distribution

For payment-free content, the transactional model is simpler than that described in the previous section. There is no financial clearinghouse. The usage clearinghouse is optional. If present, its role is mainly to monitor usage of content by consumers.

In the case of Enterprise DRM, the access control manager may implement a policy server. The policy server, after authenticating the user, defines the usage rights depending on the corresponding role and associated policies [188]. It may, for instance, use some role-based policies [189] (Fig. 4.5).

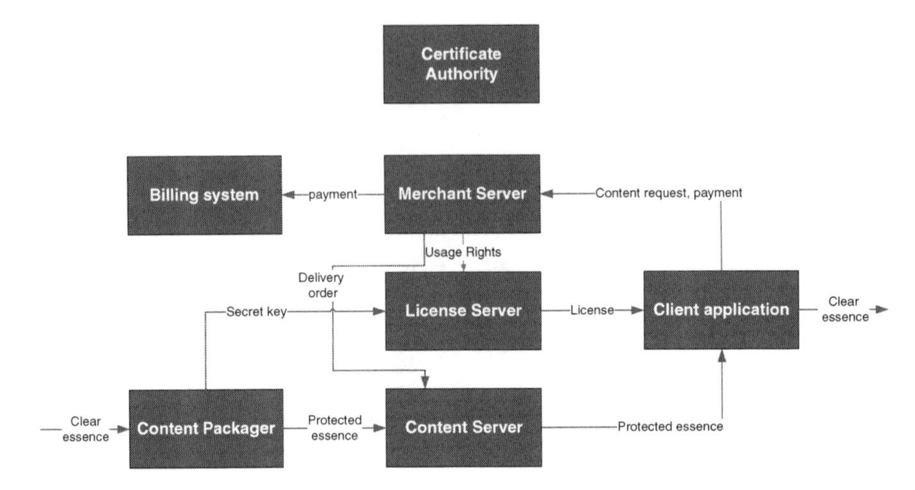

Fig. 4.6 Typical components of DRM

A usage clearinghouse is also important in a non-commercial environment such as an enterprise. It allows auditing an end user's access to protected content. Many companies may be interested in monitoring the use of their confidential information [190] to prevent data leakage. This is often done with prior notification to the end users of such monitoring to comply with privacy legislation. Regardless of legislation, the awareness of being monitored is often a strong deterrent strategy. Fear of being caught is often a good defense.

4.4 Architectural Model

The previous sections presented different models, but did not map these models to actual physical implementations. That is the goal of this section. Figure 4.6 describes a typical DRM architecture, presenting the actual hardware components of the system. A typical DRM architecture has at least the following components [191]:

- A content packager is software that prepares the content for distribution.
- The content server delivers the protected essence.
- The merchant server sells the content to a customer.
- The license server, under the control of the merchant server, delivers the license to the client application.
- The certificate authority generates for each principal of the system the secret data needed to provide authentication. It is often based on asymmetric cryptography and uses signed certificates [192].
- The client application renders the clear essence to the customer if authorized.

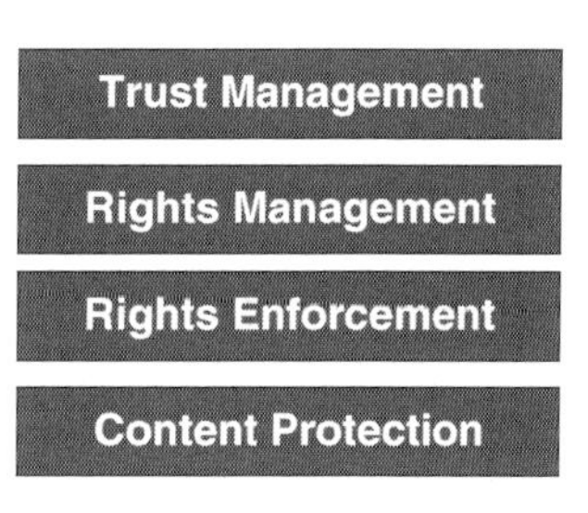

Fig. 4.7 The four-layer model

Let us map these components with the transactional model (Sect. 4.3). The distributor manages the content packager, content server, and license server. The merchant server corresponds to the access control manager. The billing system corresponds to the financial clearinghouse.

The content packager receives from content providers the content that it will distribute on their behalf. In most current implementations, this delivery is in the clear. This is of course a point of weakness vulnerable to insiders' attackers. The next generation of packagers will receive content through secure channels using encryption, forensic watermarking, or even B2B-dedicated DRM. The content packager converts the content to a format that is consumer friendly, i.e., one that the client application supports. It may encompass audio/video format conversion, repurposing, and, eventually, encryption.

Encryption may occur in one of two places: the content packager or the content server. The advantage of scrambling at the content packager is that stored content is always scrambled. This limits the risk of leakage within the DRM premises.

4.5 The Four-Layer Model

Previous models are not sufficient for fully modeling a DRM system. Presenting a communication system in a layered model is common practice in information systems. The most known layered model is the seven-layer OSI model for communication. Pramod Jamkhedar and Gregory Heileman propose a layered model of DRM compatible with the OSI model [193]. Their model uses five layers. The two upper layers fit with the actual OSI layers. In their latest work, they introduce a lower layer, the physical layer [194]. Although they attempt to tackle some security issues, they do not go far enough (Fig. 4.7).

We propose a four-layer model that is mainly security-oriented. The layered model describes the security of a DRM or a content protection system through four main features: content protection, rights enforcement, rights management, and trust management [195]. We will use this model in this book to describe many commercially deployed systems.

The content protection layer is the lower layer of this model. It securely seals the content so that no attacker may access the piece of content without having the right information. The typical security mechanism is encryption. The content protection layer on the server side scrambles the essence of the content using a secret key and a defined scrambling algorithm.[4] This algorithm is often referred to as bulk encryption. The outcome is sometimes called a secure container. The content protection layer on the client side may descramble the essence if it receives the secret key from the rights enforcement layer. The content protection layer may also alter the content to possibly trace the consumer's identity using forensic marking. The alteration may be visible or invisible. For instance, the content protection layer may append a logo or the viewer's name to the essence. It may also embed an invisible watermark with the identity of the player (Sect. 3.3.3, Forensic Marking).

The rights enforcement layer ensures that the DRM will obey the expected commercial rights. Thus, it has two main roles. First, it protects the usage rights associated with the content against tampering. An attacker should not be able to modify the usage rights. Second, it guarantees that the usage rights are complied with and not bypassed. An attacker should not be able to use the piece of content in a way not authorized by the usage rights. On the server side, the rights enforcement layer encapsulates the usage rights defined by the rights management layer, the secret key used by the content protection layer, and some ancillary data into a data structure. We will use the generic term of license to describe the corresponding data structure. It may have different names in commercial applications such as Entitlement Control Message (ECM), rights object, or Digibox. The license is (or at least some of its elements) signed and encrypted. Signature prevents tampering of the usage rights. Encryption prevents eavesdropping on the transfer of the secret key. On the client side, the rights enforcement layer decrypts the license and checks its signature. Together with the rights management layer, it decides if the user is authorized to access the content. If so, it forwards the secret key to the content protection layer.

The rights management layer handles the usage rights associated with one piece of content. These usage rights are expressed in one language. Regular Pay TV uses a rather basic language. The rights are often just Boolean access rights. Modern DRMs use more complex usage rights, requiring complex syntactic languages called Rights Expression Languages (Sect. 3.7). On the server side, the rights management layer receives from the financial clearinghouse the definition of the commercial rights granted to the piece of content. The rights management layer translates these commercial rights into usage rights using the defined language. It forwards this information to the rights enforcement layer, which will encapsulate it into the license. On the client side, the rights management layer receives the usage rights and checks whether the consumer is authorized to perform the requested

[4] The use of terms such as scrambling and descrambling to refer to bulk encryption is inherited from the early Pay TV systems. When content was analog, content was scrambled using different techniques such as variable time delay, line cut and rotate or line shuffling. Often the associated licenses were digitally encrypted. Thus, the vocabulary practice generalized for scrambling content and encrypting license. See Sect. 6.2: Pay TV: the Ancestor.

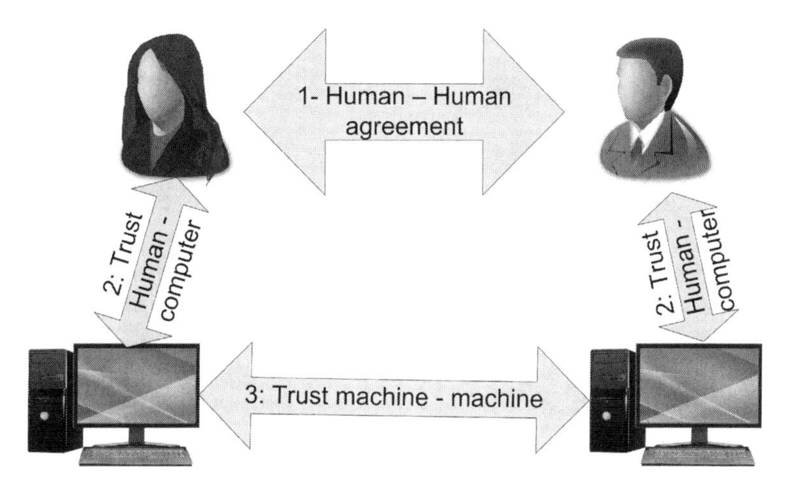

Fig. 4.8 Trust relationship

actions on the content. The rights management layer forwards its decision to the rights enforcement layer.

Trust is the cornerstone of any DRM system. Establishing trust relationships is complex [196]. It turns human–human agreements into machine–machine enforcement as illustrated in Fig. 4.8. This is the role of the trust management layer, which has mainly two goals:

- It ensures that each principal is what it claims to be, and that it is authorized to participate in the system.
- It ensures that each principal of the system behaves as expected.

The trust management layer ensures that only trustful principals interact. The system should deny any interaction between non-trusted principals or provide limited services to non-trusted principals. Trust is often "materialized" through authentication and certificates. One or more authorities issue signed certificates for all compliant principals. The trust layer checks whether the certificates are valid, and not revoked. Nevertheless, other methods are possible. Some systems use common secrets (either symmetric keys or secret functions). In any case, only compliant devices have the secret information (private key associated with the public key signed certificate, shared secret key, secret shared function) needed to interact with the other principals of the system.

The second goal is reached through legal agreements for the compliance and robustness rules (Sect. 3.8). A signature for the C&R regime implies acceptance of the rules, warranting trustful behavior.

It should be noted that the trust model of DRM is different from the traditional trust model of communication systems. This traditional trust model assumes that Alice and Bob trust each other. The trust model of DRM cannot assume this symmetry of trust. In most cases, Alice, the content distributor, cannot trust Bob,

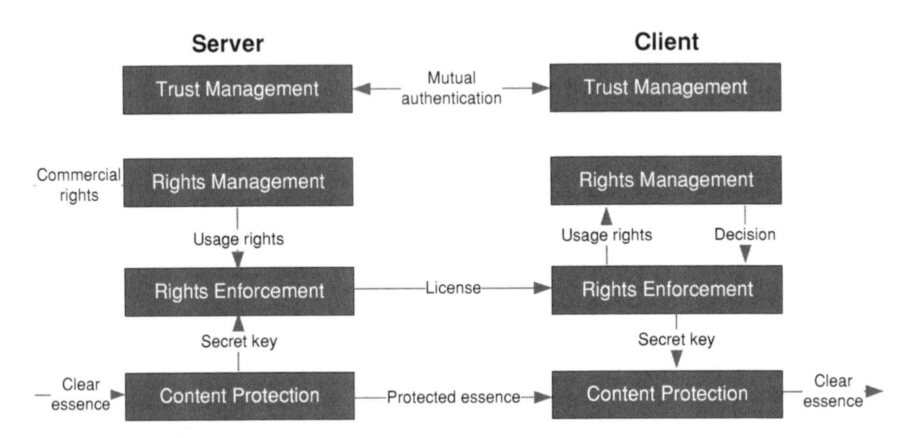

Fig. 4.9 Interaction between layers

the content user. In fact, Alice cannot distinguish an honest user from a dishonest user. Even if Bob is honest, it does not mean that his computer has not been tampered by the evil Eve [197]. This difference between the two trust models explains why content protection systems need to rely on tamper-resistance techniques (Sects. 3.5 and 3.6). Alice cannot trust Bob, but at least she may trust the tamper-resistant client application executing on Bob's computer. Section 12.1 details this difference.

Figure 4.9 illustrates the interactions of the four layers.

The Devil's in the Details

Designing correctly the Rights Enforcement Layer is one of the toughest challenges of DRM design. To illustrate the complexity of the task, we take a basic example: a monthly subscription that should work in an untethered way, i.e., the customer may access all the pieces of content once she/he paid her/his monthly subscription even if his/her set top-box cannot connect to a remote back-office server.

The Rights Enforcement Layer needs to check if the customer has a valid subscription. Thus, the Rights Enforcement Layer has to maintain on the client side a status of the entitlement. There are mainly two solutions, positive or negative addressing:

- Positive addressing means that the client application stores the status together with the end date. Before the end date, if the customer has paid his/her subscription, the license server has to issue a new license. When receiving this license, the client application updates the stored end date to reflect the new duration. The Rights Enforcement Layer checks that the status is OK and that its current date does not exceed the stored end date.
- Negative addressing means that the client application only stores the status. The server has to send a license only when this status changes. The server will send a disabling license only if the customer did not honor

his/her monthly subscription. The Rights Enforcement Layer only checks the stored status. The current date is not used.

In the early 1990s, at the birth of Pay TV, bandwidth for control messages was scarce. Thus, negative addressing was favored since it required few messages if churn of subscribers was low. Churn means that a subscriber terminates his/her subscription. The system had just to issue a message for the terminating (or new) subscribers. In contrast, positive addressing required a wealth of control messages in this case. The system periodically had to issue messages to all remaining subscribers.[5] Thus, some systems, such as Videocrypt, used negative addressing.
Videocrypt's smart card held the status of the entitlement of its owner. When receiving an Entitlement Management Message (EMM), i.e., a type of license (Sect. 6.3, How Pay TV Works) that updates the status, the smart card overwrote its status memory. Soon, crackers designed an interesting, unsophisticated attack [198]. In the early 1990s, smart cards needed two power supplies. V_{cc} was the regular 5 V power supply for the card. To write data to nonvolatile memory, the card needed an additional, higher, programming power supply: V_{pp}. Depending on the memory technology used by the smart card, V_{pp} was in the range of 12–24 V. Without the presence of V_{pp}, data could not be physically written to Electrical Erasable Programmable Read Only Memory (EEPROM), despite programs attempting to write to it. Thus, attackers simply disconnected V_{pp}. Once the smart card received the EMM requesting it to cancel its entitlement, the card attempted to overwrite the current entitlement status. The absence of the programming voltage V_{pp} prevented the memory from being actually modified. The smart card could never be revoked. The acquired rights were permanent.
A few years later, new generation of smart cards did not require this special programming voltage. Internal circuitry, using an on-chip oscillator and a diode/capacitor network, generated directly from normal V_{cc} the required higher programming voltage. Therefore, current smart cards do not any more need V_{pp}. This technological progress outdated this attack.
Another possible attack is filtering the EMM requesting cancellation of the entitlement. In the early Pay TV years, this attack was called the "Kentucky Fried Chip" [199]. A small system monitored the communication between the set top box and its smart card. It filtered out all the EMMs when detecting them. Thus, it was vital for preventing filtering. Some techniques are available. For instance, the card may expect at least one EMM every *n* seconds. If the card does not receive an EMM in this period, then the

[5] In Pay TV, the problem is exacerbated. The license server has to issue several times the same license because it does not know if the targeted set top box is powered on. Only repetitions of the message may reasonably ensure that the set top box received the renewed license.

card becomes aware that someone is cheating. Of course, the attacker may attempt to only send harmless EMMs. As a countermeasure, the EMMs will have an increasing freshness index. Each new EMM has an index value greater than its predecessor. If the card detects that the freshness index does not increase or regress, then it becomes aware that someone is cheating. Once cheating detected, the smart card may deactivate itself or take other preventive measures.
Lesson: If the security of a system relies on an important data being written to non-volatile memory (in fact, to any type of memory), then the system should actually check that this data was effectively written to memory. Furthermore, it should have a fallback solution if eventually the modification fails. This is now part of the usual best practices of smart card programmers.

We will map the architectural model to the four-layer model. The content packager and content server implement the content protection layer of the layered model. The license server implements the rights enforcement layer, whereas the merchant server implements the rights management layer. The client application implements all four layers.

Chapter 5
The Current Video Ecosystem

In the early days of TV, there was one source of video programs coming into the home: public broadcast TV. In 1929, Germany became the first country to broadcast a TV program, quickly followed by other European countries and the US. Soon the number of broadcast channels increased. In the 1970s, a second source of video program appeared through the advent of video tapes (VHS or Betamax). At the end of the 1980s, the first digital broadcast programs started. MPEG2 allowed the transition from analog to digital. Today, there are multiple sources of video content for the home. People have access to a huge variety of programming choices though their coaxial cable, satellite dish, DVD, or broadband connection. As illustrated by Fig. 5.1, there are mainly four means of carrying copyrighted video to the home:

- broadcast networks
- broadband networks
- mobile networks
- pre-recorded media

Broadcasters deliver content through broadcast channels. Broadcast channel implies that the same program is transmitted simultaneously to every viewer. Broadcast channels are one-way: information only flows from content providers to viewers. There is no return channel, i.e., no information flows back from viewers to content providers. Broadcasters may use terrestrial, cable, or satellite carriers. There are mainly two types of broadcast channels:

- Free to Air (FTA) channels provide the content for free to everybody. The reception of the signal is done through a tuner. A tuner is an electronic function which extracts the video signal that was carried, in a modulated form, by terrestrial, cable, or satellite signal carriers. Usually, the terrestrial tuner is integrated into the TV set. Cable and satellite tuners are usually implemented in external devices, so-called decoders that integrate the tuner and the decoder. The reception of terrestrial digital TV initially required an external decoder until the advent of a new generation of TV sets integrated the digital tuner. With the

E. Diehl, *Securing Digital Video*, DOI: 10.1007/978-3-642-17345-5_5,

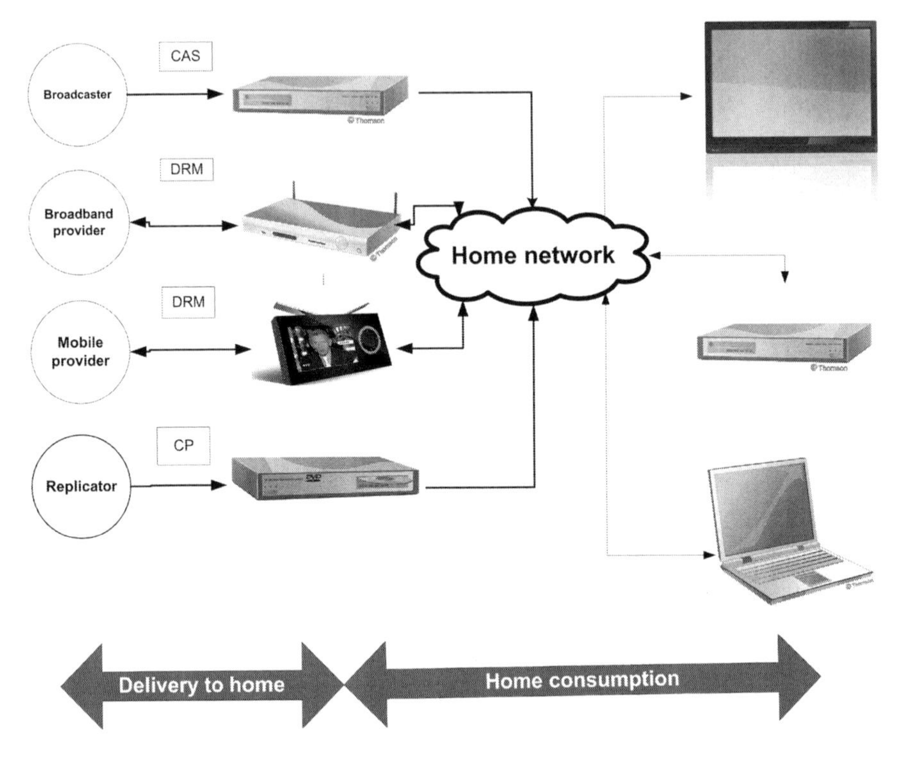

Fig. 5.1 Delivering video to the home

sunset on analog TV, all TV sets will host a digital tuner. Many countries have already enforced or will enforce through regulation this analog end in the coming years.[1]

A broadcast program is usually not protected during transmission. Broadcasters acquire the commercial rights of content only for their local market. A French broadcaster will purchase the rights of a TV show only for its French viewers. Thus, national broadcasters may have to restrict the access to free channels only to their national viewers. Else, the broadcaster may have to renegotiate rights that are more expensive for a potentially larger audience. This geographical restriction is not an issue for terrestrial transmission, since only foreign people living near the frontier may receive the terrestrial signal.[2] This is different in the case of satellite transmission; the footprint of the satellite spreads far beyond national frontiers. To enforce the negotiated commercial rights, satellite broadcasters may have to use a kind of Conditional Access System (CAS) to

[1] In 2006, Luxembourg was the first country to end analog TV. Many European countries have already followed. Since 2009, it has already been the case in the US. The US was the first non-European country to perform the switchover.

[2] Terrestrial transmission has limited range. Any physical obstacle blocks the terrestrial signal.

restrict the potential audience to national viewers. The US broadcast flag (Sect. 6.1) is a special case where US regulations mandate manufacturers to enforce copy protection of free to air content once received through the digital tuner. Of course, cable transmission does not have this footprint issue. In this case, the broadcaster precisely controls its distribution through its direct connection.

- Pay TV channels provide content only to their paying subscribers. The reception of Pay TV channels requires a dedicated device called Set Top Box (STB). Traditionally, a CAS protects content from piracy (Sect. 6.3, How Pay TV Works). Initially, Pay TV systems were just unidirectional. Later, they used phone lines as return channels to send back information, such as the actual consumption by a subscriber, to the subscriber management system. In the case of distribution via cable, CAS may use the upstream channel to carry back the information.[3]

Historically, the second source of distributed video has been pre-recorded physical media. In the 1970s, the first recordable video tapes appeared. In 1975, Sony introduced the Betamax format. In 1976, JVC, the Japanese Victor Company, introduced the rival Video Home System (VHS) format. A long battle of formats started, often referred to as Betamax versus VHS. Although the video quality of Betamax was reckoned superior to VHS's, VHS won the battle. The main reason for VHS's victory may have been that it was a royalty-free open format. Hollywood also preferred the VHS format because VHS tapes could run for two hours, compared to the one-hour length of Betamax. The two-hour duration was ideal for the new market of video home rental. It was possible to package a complete movie on one unique tape. Packaging a movie on two tapes, as would have been the case with Betamax, would have been more expensive to manufacture and would have been less user-friendly to consumers. Soon, video tape recorders became mainstream. Ant piracy appeared. Ordinary people made copies for their friends and neighbors, easily circumventing a first content protection system: Macrovision Analog Protection System (APS). Nevertheless, ant piracy was not much an issue for Hollywood. Analog copy impairs video quality. As illustrated by Fig. 5.2, successive copies of an analog tape had decreasing video quality. Although ant piracy was not an issue, large-scale piracy was a serious issue. Soon, illegal manufacturers flooded the market with illegal copies replicated in high volumes, aka bootlegging. Fortunately, bootleg piracy required serious investment in factories and infrastructure, limiting its spread. Legal enforcement was the usual answer to bootleg piracy.

In 1995, the DVD forum, founded by JVC, Hitachi, Matsushita, Mitsubishi, Pioneer, Philips, Sony, Thomson, Time Warner, and Toshiba, introduced the DVD. DVD-Video offered superior quality and a smaller form factor than VHS. The CD audio had already familiarized customers with this new form factor.

[3] Cable transmission is bidirectional. The downstream link delivers data to the home whereas the upstream link carries data from the home to the cable head end.

Fig. 5.2 Analog copy versus digital copy

Although expensive at the beginning, recordable DVDs were available at the same time that DVD players were introduced in the market. Unlike analog copy, digital copy is pristine. Successive copies of a DVD do not impair the quality. Hollywood had learned from the audio industry about digital piracy. When the DVD was first introduced, ripping CD audios was already common practice. CD audio is not protected against duplication [200]. Though audio compression such as mp3 was not yet designed in 1995, audio piracy was already ramping up. Thus, studios decided that DVDs should be protected against duplication by consumers. Therefore, CSS protects DVD content (Sect. 8.2).

Ten years later, high definition video became accessible. Large Liquid Crystal Displays (LCD) or plasma flat screens were overthrowing traditional analog TV screens. The DVD format did not support HD video. Thus, a new format of pre-recorded media was necessary. A short format battle started between the Blu-ray and the HD-DVD. In 2007, the HD-DVD consortium gave up. Today, there is one single format of HD content: Blu-ray. The Advanced Access Content System (AACS) and BD+ protect Blu-ray discs (Sect. 8.5, Blu-ray).

At the same time, high-speed broadband connection became ubiquitous. Bandwidth constantly increases, at a rate faster than that of Moore's law. Delivering high-quality video through a broadband connection was not anymore a dream. Broadband providers can deliver HD programs through Internet two-way connections. The content is delivered through the downstream but the protocols use the upstream to attain better quality (through Quality of Service (QoS)) or to implement additional services for the consumers and security features. Broadcasters use unicast or multicast models [201] to deliver their broadcast programs via IP connections. In the case of unicast, the content server sends the content to one unique receiver. Each emitted IP packet is dedicated to one unique given IP address. In the case of multicast, the piece of content is sent simultaneously to several receivers. A multicast IP packet is sent to a set of IP addresses that are part of a multicast group. To receive content, the broadcaster provides a dedicated gateway to its customer. Current gateways host many functions such as modems, routers, and firewalls. Furthermore, often the gateway acts also as a wireless access point. Modern gateways support triple play offer (data, voice over IP, and video). Traditionally, DRM systems protected video content being transferred over the Internet (Sect. 7.1, DRM). There are mainly two types of controlled video distribution:

- Network Service Provider (NSP) or Internet Service Provider (ISP) commercial offers, where the operator has a website, the portal, where customers may purchase VOD content. The operator also distributes broadcast TV channels. The competitive advantage of the NSP is that it controls the distribution channel. It can thus control the use of the bandwidth of its network to guarantee QoS.
- Over The Top (OTT) commercial services, where the content is distributed independently from the ISP's control. It is a "normal" connection to a Web merchant site. In this case, it is more difficult to guarantee QoS because the video server has no control over the allocated bandwidth. NetFlix is an example of such an OTT service.

The latest deployed video distribution channel is the mobile one. Telecom mobile operators deliver content to mobile phones or mobile appliances such as the iPad, for instance, through 3G or Universal Mobile Telecommunication System (UMTS). The connection is bidirectional. Of course, the receiver is a mobile phone with a video display. Dedicated DRM systems protect content (Sect. 7.5, Open Mobile Alliance).

Today, most content providers use all available means to distribute their content, sometimes even in a combined supplementary offer. For instance, a TV broadcaster may offer its show first through so-called linear TV, i.e., typical broadcast channels and TV channels on broadband connections. Once a show has finished, it will be available as catch-up TV, i.e., a kind of VOD, after a delay of one hour to one week. Viewers can visit the website hosting the catch-up TV and watch the streamed show at their convenience. In the case of Pay TV program, the access to catch-up TV is limited to subscribers. After a while, the show may be available as DVD or Blu-ray. In other words, the show has been versioned to maximize revenues. Catch-up TV increases advertisement revenues whereas DVD increases sales.

The future trend is in the digital home. The digital home will connect all devices in a seamless way. Already the new generation of TV set is directly connected to the Internet. Consumer will be able to view any content on any device of the home regardless of its source. The home network encompasses fixed devices and mobile devices. The digital home requires a different set of protection technologies described in Chap. 9.

We must not forget that the first historical source of video was cinema. For more than a century, cinema has used celluloid reels. With reels being physical analog goods, security was not an issue. In the 1980s, digital versions of movie were appearing in postproduction as digital intermediates. In 2000, "Chicken Run" was the first movie fully digitally processed. In 1987, Texas Instruments introduced the Digital Light Processing (DLP) technology that would enable digital projection. The first digital projector appeared in the 1990s with quality

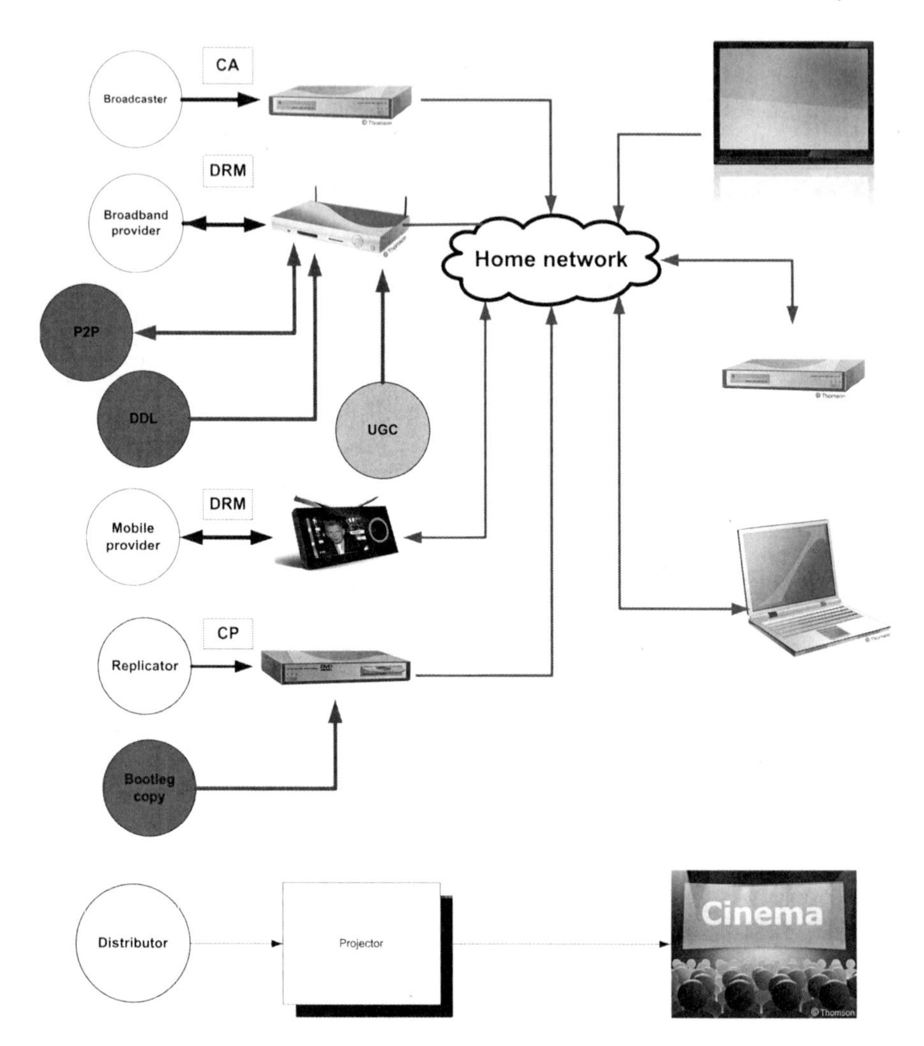

Fig. 5.3 The real-world video context

poorer than traditional 35 mm projectors. Current digital projectors have high quality and high resolution, typically 2 K (2,048 × 1,080) and even 4 K for the high end. Since 2005, with the completion of standards (Chap. 10), digital cinema started to spread and to replace 35 mm projectors.

Figure 5.1 describes an ideal world. Figure 5.3 provides a more realistic vision of the distribution of video content to the home. There are additional sources of copyrighted content. An older alternate source is the bootleg, in other words, the pirated DVDs. Obviously, these DVDs are not protected by any protection system. As is the case for genuine DVD, the market for pirated DVDs, at least in Western countries, is shrinking. Currently, the main alternate sources are P2P and Direct

Download (DDL) sites. Most of the pieces of content available on P2P and DDL are illegal. The last alternate sources are the UGC sites such as YouTube and DailyMotion. On these sites, most of the proposed contents are generated by users and thus not protected by any copyright. Unfortunately, some users post copyrighted content. Most UGC sites have a process to address this issue [61]. As soon as a content owner notifies a UGC site of the presence of its copyrighted content, the UGC site removes the allegedly illegal content. For more details see Sect. 2.4, A Brief Overview of Digital Piracy.

Chapter 6
Protection in Broadcast

6.1 The Broadcast Flag

With the advent of the Advanced Television Systems Committee (ATSC), the new US digital terrestrial broadcast system, US studios and broadcasters were afraid of a potential rise of piracy. They feared that their broadcast content could be too easily redistributed over the Internet and this may compromise their advertising revenues. In November 2003, after strong lobbying, the Federal Communications Commission (FCC) adopted the broadcast flag regulations with the goal of preventing indiscriminate redistribution of content over the Internet or through similar means. The announced timeline for the broadcast flag was for it to be active on June 5, 2005.

In the US, FTA television, non-premium subscription channels, and "basic" cable channels have to be broadcast in the clear and available for copying as backup. Thus, it was impossible with regulation to protect these programs through encryption while they were being broadcast. The FCC could not legally allow encryption during transmission. In May 2001, the Fox Broadcasting company presented to ATSC a technical proposal for the broadcast flag. Following this presentation, the subcommittee Broadcast Protection Discussion Group (BPDG) of the Copy Protection Technical Working Group (CPTWG) supported this proposal. The broadcast flag is one bit of information inside the ATSC bitstream. It was defined as the Redistribution Control Descriptor flag. This flag is present in the digital program guide tables and also in the Program Map Table. Once received by the demodulator, i.e., the digital tuner, the clear content has to be protected when the broadcast flag is switched on (Fig. 6.1). When processing protected programs, the digital output of the receiving device should use an approved copy protection technology: an IP connection to a Wide Area Network (WAN), Local Area Network (LAN), or any other digital bus such as IEEE1394 or HDMI. If the receiver features a digital recorder, it should ban copying if the copy cannot be restricted to personal usage. Similarly, analog output of HD content should be downsized to SD.

E. Diehl, *Securing Digital Video*, DOI: 10.1007/978-3-642-17345-5_6,

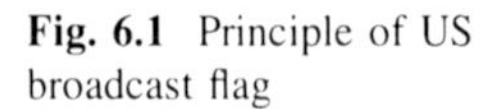
Fig. 6.1 Principle of US broadcast flag

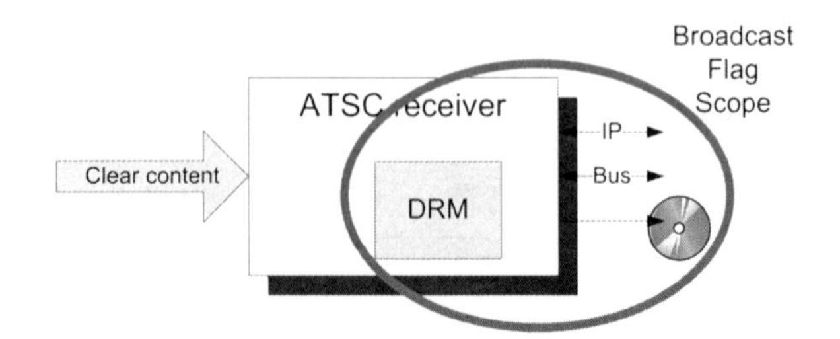

A compliant ATSC receiver could export broadcast flagged programs only through approved technologies listed in the so-called Table A. The FCC had the responsibility of deciding which technologies were suitable for Table A. In August 2004, FCC approved thirteen technologies, among them CPRM (Sect. 8.4), D-VHS, DTCP (Sect. 9.3), HDCP (Sect. 9.4), Helix DRM, Magic Gate (Sect. 8.3), Microsoft DRM (Sect. 7.2), SmartRight [202], Tivo, and Vidi Recordable DVD Protection System [203].

The Redistribution Control Descriptor flag was in the clear in the bitstream and not protected for integrity. Simple digital circuitry or simple software could easily overwrite the flag. Nevertheless, the unique goal of the broadcast flag was to enable legal action through DMCA in the case of circumvention. Thus, robust protection of the broadcast flag was useless. Nevertheless, the broadcast flag made sense only if mandated by regulatory requirements, and ditto for the FCC ruling. The combination of FCC ruling and DMCA allowed such legal action [17]. As such, the broadcast flag was similar to a hook IP, although it was not built on patents but only on regulations (Sect. 3.8, Compliance and Robustness Rules).

On May 6, 2005, the US Court of Appeals for the District of Columbia Circuit struck down the FCC's broadcast flag rules [204], holding that the FCC lacked jurisdiction to regulate how televisions and other devices should handle content after reception of broadcast transmission. The current official mandate of FCC only covers the reception of broadcast signal but cannot extend beyond reception. This decision signed the death of the broadcast flag. The broadcast flag was never deployed.

6.2 Pay TV: The Ancestor

Some authors trace the first forms of DRM to the advent of mini servers and mainframes in the 1970s [191]. In my opinion, this is not true, or at least not true for multimedia DRM. These early systems lacked one key component of efficient content protection: encryption. At their time, cryptography was reserved for defense applications. In my opinion, because they rely on cryptography, Pay TV systems are the first form of video content protection and DRM

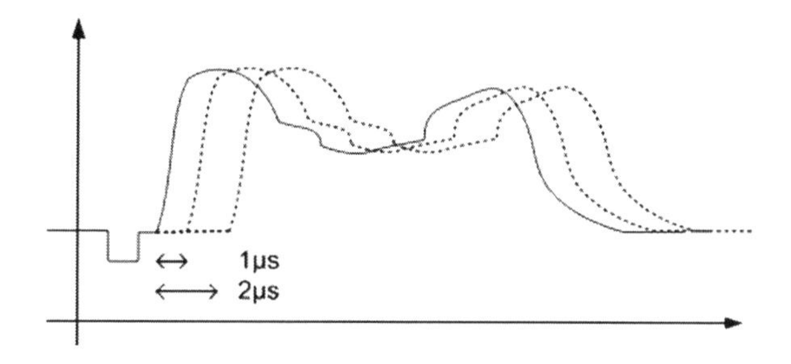

Fig. 6.2 Analog scrambling principle of Canal+

Pay TV is often compared to FTA TV.[1] Both systems broadcast their programs to consumers. Pay TV delivers broadcast programs only to paying customers, whereas FTA delivers programs to all customers. The security objective of Pay TV is to ensure that nobody can view Pay TV programs without having paid for it. Pay TV uses the same carrying signal mechanisms and media as FTA. This means that the broadcast signal is available to anybody. This signal is non-excludable (Sect. 2.3). Therefore, it is mandatory to create artificial excludability to restrict access to the carried image. This is the role of CAS, the first instance of video content protection.

In the early 1980s, audio and video signals were not protected. Radio broadcast was free, and audio tapes were not protected. There only were FTA operators, and video tapes were not protected. Thus, the history of content protection started with the advent of Pay TV channels. In 1984, the French operator Canal+ launched its first subscription-based channel. The video signal was analog, and the transmission was terrestrial. The protection principles were rather simple. Content was scrambled, i.e., content was transformed to become non-viewable. The received analog signal was digitized. These new digital technologies allowed key-based scrambling techniques.

For instance, Canal+ inserted a variable delay after the synchronization signal of each video line. Figure 6.2 illustrates the scrambled video. An algorithm generated a pseudo-random suite of 1,800 values in the set {0, 1, 2}. Depending on the random value for the current line, the scrambler added 0, 1, or 2 µs after the video synchronization pulse signal. This resulted on an extremely jittery image. Once per month, the subscriber received an activation code. Once the user dialed the right activation code in the decoder, the decoder could generate the right pseudo-random suite of 1,800 values. With this suite of values, the decoder could descramble the video signal, i.e., the decoder could reverse the applied transformation by removing the appended delay. Soon, the first schematics of pirate decoders appeared. Unfortunately, finding the short monthly key was rather easy after some trials. This was done by toggling some physical switches until the video

[1] In the US, the separation is more complex because there are also premium channels such as HBO and ShowTime.

was correctly descrambled. Later, a new version of pirate decoders that automatically guessed the monthly key appeared in the market.

Canal+'s design had several severe flaws:

- With only eleven bits, the length of the key was too short. It only took a few minutes to by brute force find the 11-bit key.
- The lifetime of this 11-bit key was too long. A key was valid for 1 month. The pirate had only to spend a few minutes each month to guess the code. Anybody was able to do it. The incurred inconvenience was minor compared to paying a monthly subscription, and thus not a deterrent enough. Many people were ready to spend a few minutes on a dumb task to save the cost of a subscription.
- The scrambling algorithm was simplistic. It was too easy to build a descrambler. Some pirate decoders even defeated the scrambling without using the pseudo-random generator. They measured the delay in the scrambled signal between the end of the synch pulse and the start of the effective video signal and compensated for it.[2]

In 1985, HBO used Videocipher II to protect its satellite programs. Soon, pirate decoders were available in the market. Pirates offered distribution channels that delivered the monthly key, as well as updates and fixes, to their customers with subscription models. In the case of Videocipher II, piracy was organized as quasi-industrial organizations [205]. The pirates were able to defeat countermeasures by HBO in a few days. It was estimated that the "market share" of pirated decoders was higher than that of HBO. The Satellite Broadcasting and Communications Association estimated that there were 3.7 million deployed satellite dishes although only 1.7 million were served by an official subscription!

These first commercial deployments of Pay TV highlighted the need for longer keys and better means to protect these keys. Two European Conditional Access systems, Eurocrypt and Videocrypt, built the foundations of modern content protection. The Union Européenne de Radiodiffusion (UER) established Eurocrypt as a standard. The Eurocrypt specifications established the vocabulary that is still in use today in Pay TV systems [206]. Videocrypt was co-designed by News Data Com (the future News Data System (NDS) group) and Thomson [207]. These systems created a new generic scheme for content protection [199].

The first improvement was that these new systems used robust analog scrambling schemes such as line cut and rotate, or line shuffling. The line cut and rotate, or active line rotation scheme, defines 256 points in the video line. For each successive video line, in fact, a pseudo-random number generator, seeded by a key, selects a random cutting point from among these 256 points. The cutting point splits the video line into two segments. Scrambling inverts the order of these two segments. Descrambling reverts the order of the two segments to their initial

---

[2] After a few years, it was a common exercise in some French electronic engineering schools to study how to hack the Canal+ signal.

Fig. 6.3 Line cut and rotate scrambling

position. The cutting point changes every successive video line. Figure 6.3 shows the corresponding visual effect.

The line-shuffling, or line permutation scheme, works with blocks of consecutive video lines. Scrambling applies a mathematical permutation to the lines of the block. Descrambling applies the reverse permutation. A pseudo-random number generator seeded by a key defines the permutation. The British Broadcasting Corporation (BBC), in collaboration with Thomson, attempted to use such a scrambling technique for its BBC Select service [208].

Although these scrambling algorithms were far more robust than traditional analog scrambling algorithms, the major improvements came from the combination of sound cryptography [209] and smart cards. Smart cards are removable secure processors (Sect. 3.5.2). In 1979, the first smart cards were commercially available. French manufacturer CII Honeywell Bull (the future Bull) proposed the first smart card CP8 [114]. The mask called PC2 (Porte Clé 2) soon became the major cornerstone of future Pay TV systems. Smart cards allowed the following:

- The use of modern cryptography for efficient key management.
- A processor dedicated to security; one specific functionality is that communication with the processor is done through a controlled limited set of commands. The processor only uses embedded volatile and non-volatile memory. There is no direct external access to the volatile or non-volatile memory of such a processor.
- Tamper-resistant hardware, which protects the memory and the processing unit. This means that keys and software are protected. Reverse engineering tamper-resistant hardware requires both more skill and more specialized, expensive equipments than reverse engineering software (Sect. 3.5).
- Renewability; an initial assumption of the trust model of modern pay TV was that crackers would ultimately break the system and especially the key management. The model assumed that the scrambling method would not be cracked. In other words, the design assumed that it was impossible to build a

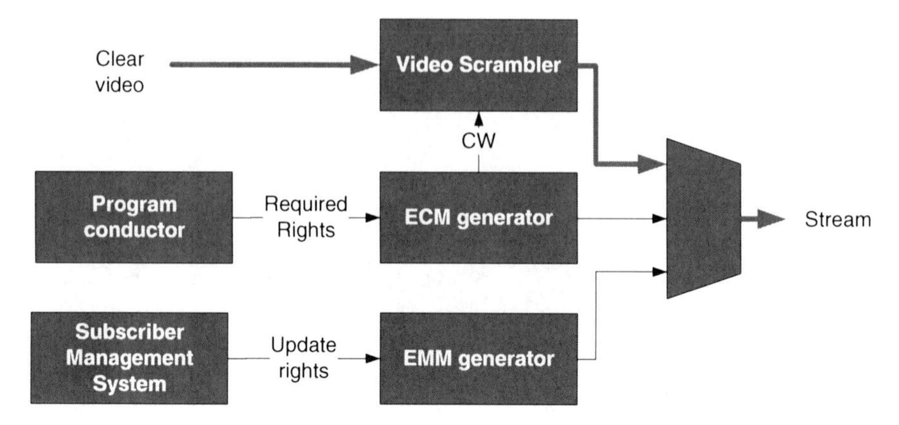

Fig. 6.4 Simplified architecture of a pay TV head-end

pirate decoder without first breaking the key management.[3] The smart card implemented the key management and the management of the entitlement rights. Changing the smart cards was a powerful way to address future potential hacks of key management. This approach has proven to be successful.

Eurocrypt and Videocrypt survived many years. They were phased out by obsolescence of the video technology rather than by crackers. Eurocrypt used D2MAC, which was phased out by new satellite technologies, whereas Videocrypt used PAL analog signals, which was phased out by digital TV.

Then digital video appeared. Video compression techniques allowed to efficiently turn analog video into reasonable size digital data. The MPEG2 compression technique became a standard. New modulation schemes allowed carrying these digital compressed videos. It was possible to stream MPEG2 compressed video to satellite set top boxes. In 1995, the Digital Video Broadcast (DVB) group standardized the way to protect MPEG2 transport streams. It defined a common scrambling algorithm, the DV Common Scrambling Algorithm (DVB-CSA), and the signaling for proprietary CASs (ETR 289 [210]).

6.3 How Pay TV Works

Figure 6.4 describes the main elements of a Pay TV head end. The video scrambler protects the clear essence by scrambling it with a symmetric encryption algorithm. The encryption key, called the Control Word (CW), is valid for a given duration

---

[3] Line cut and rotate scrambling was never broken (at least practically). One implementation of line shuffling was broken due to a weak implementation of the random permutation. This permutation used a limited table of fixed permutations that was reverse engineered. Once the table was disclosed, it was easy to build a pirate decoder. In other words, the entropy of the permutation was too small.

known as the crypto-period. Usually, crypto-periods are about a few seconds, typically about ten seconds. The scrambling is done on the fly, i.e., in real time. In fact, often both the video stream and the audio stream(s) are scrambled, each one using its own CW. In the remainder of this section, we will not differentiate video from audio.

The Devil's in the Details

The definition of the scrambling algorithm has some interesting constraints that exceed the simple definition of a cryptographic algorithm [211]. The scrambling algorithm defines the following:

- the cryptographic algorithm itself, for instance, AES; and
- the chaining mode, for instance, ECB, CBC, or CTR.
  - ECB (Electronic Code Book) is the simplest mode of encryption. As illustrated in Fig. 6.5a, the plaintext block is encrypted using the CW. It has the advantage of being simple. Unfortunately, scrambling the same block with the same CW always produces the same scrambled block. This may sometimes be an issue, for instance, if some blocks of a stream are always constant. ECB is the weakest mode.
  - CBC (Cipher Bloc Chaining) is a more complex mode. As illustrated by Fig. 6.5b, the first plaintext block of the sequence is fed with an Initialization Vector (IV) and then encrypted using the CW. The subsequent plaintext blocks are XORed[4] with the previously scrambled block and the result is encrypted using the CW. This mode does not support parallelization. Furthermore, it propagates possible errors. This may be an issue for noisy transmissions.
  - CTR (CounTeR mode) is an interesting mode. As illustrated by Fig. 6.5c, it turns a block cipher into a kind of stream cipher. Each plaintext block is XORed with the output of the encryption of a counter using the CW as the key. The counter may be a simple counter or any type of sequence that does not repeat for a long period. The advantage of CTR mode is that it allows easy random access to any position of the streamed content.
- The Initialization Vector if needed. The IV may be fixed or may be changing, in which case it will have to be carried with the license.

We described scrambling based on conventional cryptosystems. There exists another class of scrambling that uses another approach. It is called multimedia encryption or selective encryption. Encryption-based scrambling does not take into account the data structure of the compressed essence.

[4] XORed means applying the Boolean exclusive OR operator. The result of the XOR operation on two operands is true if exactly one operand is true.

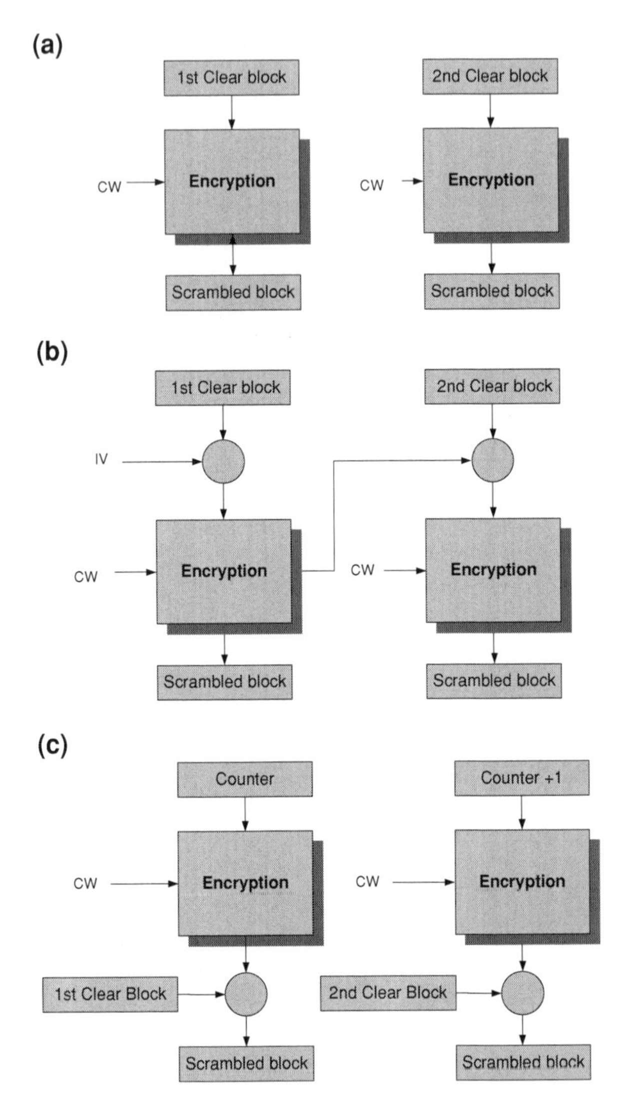

Fig. 6.5 Different crypto chaining modes. **a** ECB mode. **b** CBC mode. **c** CTR mode

It encrypts all the bytes. Selective encryption encrypts or modifies only a limited portion of the essence. It takes into account the data structure of the container and codec. One of the characteristics is that, often, selective encryption preserves the multimedia format, i.e., the syntax is respected. As a result, the scrambled content may give some visual hint of the content [212, 213].

Lesson: When discussing bulk encryption, be aware of the details of encryption modes. That two content protection systems use 128-bit AES does not mean that their content is interoperable.

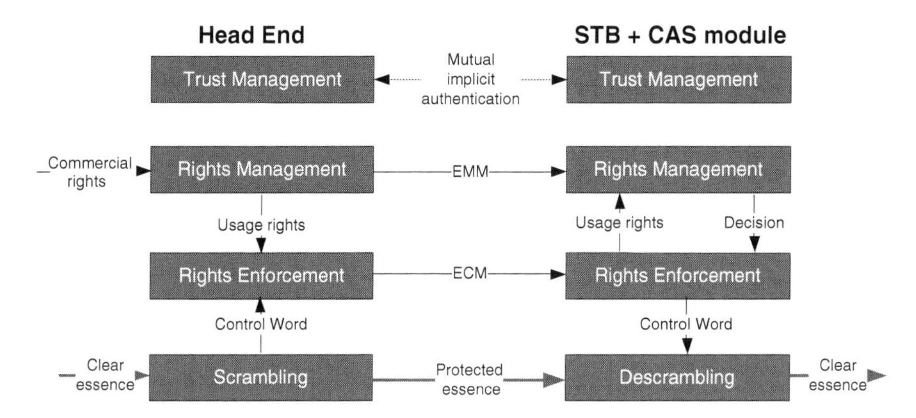

Fig. 6.6 Four-layer model of pay TV

The ECM generator randomly creates the new CW and delivers it to the scrambler. The ECM generator builds a data structure called an ECM, which will enable the descrambling of the program. An ECM contains at least the current CW, the identification of the corresponding program (program ID), and the usage rights needed to access this program. The program conductor delivers the program ID and the corresponding usage rights to the ECM generator. The program conductor has all the information of the program guide. Thus, it "knows" the set of events to be broadcast. Often, the program conductor drives the robot that loads video tapes in the case of analog play-out, or drives the video servers that hold the digital files of the programs in the case of digital play-out. The ECM generator encrypts the ECM data structure. The Subscriber Management System handles the commercial rights acquired by the customers. It passes this information to the EMM generator. The EMM generator builds an EMM that updates the rights granted to the consumer. An EMM contains the identity of the targeted customer or group of customers and the new granted usage rights. The EMM generator encrypts the EMMs. The ECMs and EMMs are encrypted using secrets dedicated to a given broadcaster (or group of broadcasters). The scrambled video, the ECM stream, and the EMM stream are multiplexed into one unique transport stream that is broadcast. Sometimes, the head end can send EMMs through other delivery channels such as phone lines. This type of EMM emission is called out-of-band. It is useful for handling unusual events such as for urgently updating a customer's rights.

We can map this architecture to the model described in Sect. 4.4. The video scrambler is the packager. The ECM and EMM generators correspond to the license server. Indeed, in Pay TV, the license is split into two data structures. The Subscriber Management System is the equivalent of the merchant server. Of course, the STB plays the role of the client application.

The STB receives the multiplexed transport stream containing the scrambled video, the scrambled audio channel(s), the ECMs, and EMMs. It extracts the different ECMs and EMMs by demultiplexing the transport stream and filtering out the EMMs that do not concern the STB. The STB has an associated CAS module. If the CAS module belongs to the right broadcaster and is not revoked, then it holds all the secrets needed to decrypt the ECMs and EMMs. This means that the trust management is implicitly successful if both the head end and the CAS modules share some secret information. When the CAS module receives an ECM, it decrypts it and then checks if viewing is authorized. For that purpose, it compares the rights, requested by an ECM, to the rights stored in its protected memory. If the stored rights fit, then the CAS module retrieves the control word and descrambles the video stream. When the CAS module receives an EMM, it decrypts it and then checks if the CAS module is the expected recipient. A CAS module has an individual identity and some optional group identities. If the CAS module is the expected recipient, it correspondingly updates the stored usage rights (Fig. 6.6).

Let us explore two usual business models of Pay TV: monthly subscription and PPV. Monthly subscription means that the customer pays a regular fee to access a commercial offer defined by the broadcaster. The commercial offer may be either an isolated channel or a bundle of channels. The description of the usage rights (i.e., the REL) is often very rudimentary but efficient in a Pay TV context. For instance, it may use one bit for each possible channel or group of channels. When Alice subscribes to a given commercial offer, the Subscriber Management System (SMS) logs this information into its database. Then, it sends to the EMM generator the identity of Alice's CAS module, together with the description of the newly acquired rights (the corresponding bit is set to 1 if she should have access to the corresponding channel or group of channels) and the period of validity (the current month). The EMM generator builds and encrypts the corresponding EMM that is then broadcast to everybody. For each channel, the program conductor provides the description of the needed rights to the ECM generator, i.e., the bit corresponding to the broadcast channel is set to 1, and the others are null. When Alice's CAS module receives an ECM for a channel she subscribed to, the stored entitlement will fit the usage rights defined by the ECM. If the current date, also part of the ECM, is still in the valid entitlement period, then the CAS module descrambles the video. Of course, for a channel that does not belong to Alice's subscription, the stored entitlement will not fit. Before the end of the month, the SMS issues a request for valid EMM for the next month to all subscribers. Anticipating this renewal prevents the rupture of services.

In the case of PPV, the scheme is slightly different. The broadcaster defines a unique event ID for each PPV program. Once Alice purchases an event in PPV, by phone or through an interactive application using a return channel, the SMS sends the identity of Alice's CAS module together with the unique event ID of the purchased event to the EMM generator. The PPV EMM contains this unique event ID. Alice's CAS module stores the unique event ID in its secure memory. The program conductor delivers to the ECM generator this unique event ID that is embedded in the PPV ECM. When receiving a PPV ECM, Alice's CAS module checks if the corresponding event ID is already registered in its secure memory. If this is the case, the CAS module descrambles the movie. The CAS module updates its memory to log that the movie has already been viewed. Thus, the CAS module can enforce that Alice does not watch the same event several times while only paying for one session.

The described mode of PPV is rather similar to current VOD implementations. Nevertheless, there is an old implementation worth describing. Sometimes the CAS module held an electronic purse that contained tokens. In that case, the program conductor delivered to the ECM generator the unique event ID together with the access price (expressed in tokens). The ECM carried this information together with the control word. If the consumer had enough tokens in his purse, then the STB asked him to confirm the purchase. If the consumer accepted the transaction, the CAS module subtracted the cost of the event from the purse, logged the transaction, and then delivered the CW to the STB for descrambling. Through a return channel, the CAS module could feed back the log file to allow the fair redeeming of the content owners. Of course, the consumer could purchase tokens to reload his purse. The reloading was triggered by a dedicated EMM. This method was useful when the return channel bandwidth was sparse. It is currently not anymore in fashion.

The Devil's in the Details

Normally, each channel is independent. Thus, each channel uses different control words. Consequently, each time the consumer tunes to another channel, the STB has to retrieve the current control word of the newly tuned channel. The duration of this retrieval directly influences the user experience. It should be as short as possible. The usual crypto-period of a CW, i.e., its period of validity, is about ten seconds. The same ECMs are repeated within the stream (about every 100 ms). Thus, the STB will need at maximum 100 ms to retrieve an ECM. We must add the processing time of the ECM by the smart card, which takes also several hundreds of milliseconds. CAS often uses a short crypto-period. In the broadcast environment, ECMs have to be often repeated in order to retrieve quickly the clear video; therefore, CAS can afford to use short keys. For one hour of content, pirates would have to crack 360 different control words. DRM usually uses one unique key for a complete movie. Therefore, DRM needs longer keys. Nevertheless, CAS currently switches to 128-bit keys.

To facilitate the transition between crypto-periods, two successive CWs are identified as even and odd keys. The idea is that during an even crypto period the stream is scrambled using the even CW, and the stream uses the odd CW for the next crypto-period. Successive even (respectively odd) CWs are different. Two bits of the transport stream indicate the parity of the current CW [214] (Table 6.1).

The STB automatically changes the CW when these bits toggle in the stream. Current SoCs automatically manage this transition. The CAS only has to store the two CWs inside the internal registers of the SoC. The SoC automatically selects the proper CW.

Table 6.1 Management of two CW

00	The transport stream is not scrambled
01	Reserved for future use
10	Even CW scrambles the transport stream packet
11	Odd CW scrambles the transport stream packet

Lesson: Changing the scrambling key within a content is a complex task. Proper synchronization between video frames and actual key scrambling is needed. Thus, many DRM systems use one unique scrambling key for a protected file or stream.

6.4 DVB

DVB is an industry-led consortium of about 250 broadcasters, manufacturers, network operators, software developers, regulatory bodies, and others in over 35 countries, committed to designing global standards for the global delivery of digital television and data services. Services using DVB standards are available on every continent, with more than 500 million DVB receivers deployed [215].

In the early 1990s, Pay TV was just starting. The landscape of Pay TV was complex. Many European broadcasters were fiercely competing. Several Conditional Access providers, such as Nagra, NDS, or Viaccess, contended for market share. Broadcasters were supporting the cost of both STB and broadcast equipment. It became rapidly clear that price reduction could only occur if there was a mass market. Thus, it was necessary to standardize as many elements as possible. In 1993, DVB was founded with the goal to standardize them. One interesting characteristic of DVB is that it is consensus-driven. Thus, an adopted solution is a finely crafted equilibrium that respects at best the interests of all stakeholders, ensuring a higher chance of market adoption. DVB standardized the format of the video, selecting MPEG2. DVB defines the way to signal on air programs (DVB SI) [216] and the way to carry the data within the transport stream. This part of the work is not related to security issues.

On behalf of DVB, a group of international experts defined the DVB Common Scrambling Algorithm (DVB-CSA) used to protect content. DVB-CSA uses a 40-bit key Control Word.[5] DVB-CSA's crypto period can vary from 10 to 120 s. The details of the algorithms were confidential. Nevertheless, after more than

[5] In fact, DVB-CSA used a 56-bit key with only 40 bits of entropy. This was to comply with limiting regulations. Due to the short lifetime of the key, it is assumed to be long enough. A new version called CSA V2 uses the full range of 56 bits. A newer version CSA V3, using a 128 bit key, should be deployed in the coming years [217].

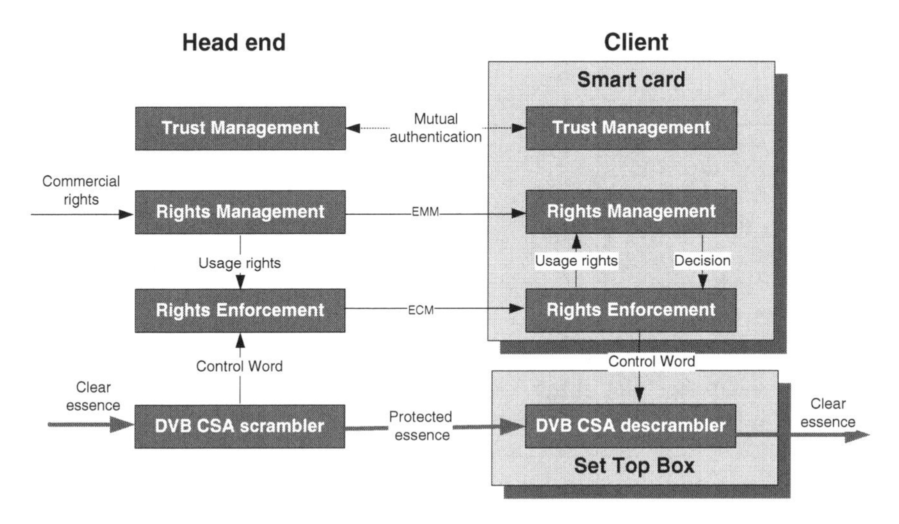

Fig. 6.7 DVB four-layer model

10 years of cover, they leaked out. Currently, there is no known practical attack on this algorithm [218, 219].

In the DVB architecture, each STB has an associated smart card. The smart card hosts the rights enforcement layer, the rights management layer, and the trust layer, whereas the STB hosts the protection layer. This architecture has the advantage that renewing these three upper layers is easy by simply updating or replacing the smart card. In fact, DVB only specifies the content protection layer, the identification of ECMs and EMMs, their size, and the way to carry them within a DVB transport stream. The actual behavior, specifications, and security of the smart card are fully proprietary for each CAS provider. In other words, each CAS may use its own implementation of the rights enforcement layer, the rights management layer, and the trust layer (Fig. 6.7).

Using a common scrambling algorithm has the advantage of lowering the entry barrier for new CAS vendors, and (in theory) allowing the broadcasters to easily swap it for another CAS if ever there is a successful hack. Furthermore, sharing the same hardware components would reduce the cost of the STB.[6]

The Devil's in the Details
In 1995, US broadcaster DirecTV deployed a smart card, designed by Nagra, to protect its Pay TV channels. Unfortunately, the designers of the smart card, called the H card, reused a design already used in Europe. This European design had been already hacked. Thus, US crackers had access to

[6] Currently, STMicroelectronics and Broadcom, two IC manufacturers, dominate the market of SoC dedicated to STB.

precious knowledge collected previously by European hackers. Good knowledge of the smart card and part of its software helped the hacking community to quickly reverse-engineer the system. Soon, they designed a card writer that could modify the subscription rights managed by the smart card.

The arms race with the hackers had started. DirecTV designed a mechanism that allowed the STB to detect hacked cards and to block them by forcing an update. The hacking community soon delivered another piece of hardware that healed the updated blocked cards. Later, the hacking community wrote a piece of software that definitively disabled the ability of DirecTV to update the pirated smart cards. Then, DirecTV launched some software patches that required the updates to be present in the card for the decoder to work. Each time, within a few hours, the hacking community released pieces of software that overcame the new countermeasures.

After a while, DirecTV launched one new set of successive updates. Surprisingly, the source code of these updates was meaningless and useless. The bytes had just to be present in the smart card. They had no visible effect. It was like garbage. The hacking community adapted the software of the hacked cards to accommodate this seemingly stupid evolution. After about thirty updates, the uploaded bytes turned the bytes already present into a dynamic programming capability.

On Sunday, 21 January 2001, DirecTV fired back. The last update checked the presence of some bytes that would only be present in hacked smart cards (we would call that the signature or fingerprint of hacked cards). If a hacked smart card was detected, the newly built software updated the boot section and wrote four bytes to the Programmable Read Only Memory (PROM) area. Among them, the value of memory 0x8000 was set to 0x00. The smart card's boot checked if the value of memory 0x8000 was 0x33 (as in genuine cards). If it was the case, the program continued and sent the normal Answer-To-Reset (ATR). Else, the program stayed blocked in an infinite loop. The hacked smart cards no more answered to the STB or to any card reader. The checked byte was in the PROM area; thus, it was not possible to reverse it to the expected value. The game was over.

This was the Black Sunday. Thousands of hacked cards were definitively dead. One interesting fact is the date chosen by DirecTV to strike back: one week before the Super Bowl (one of the biggest US TV events) [220]. Football fans had no alternative. They had to pay DirecTV to watch their annual, favorite event. The smart choice of the date was part of the success of the counterstrike.

Lesson: "Si vis pacem, para bellum".[7] This is key advice to any DRM designer.

[7] "Who wants peace, prepares for war."

6.5 DVB-CI/CI+

DVB standardizes the way to protect broadcast digital content using the DVB-CSA bulk encryption (Sect. 6.4, DVB). In this model, the proprietary part of the protection system is "exported" to a remote smart card. This model was suitable for Set Top Boxes managing one single CAS provider. It did not fit well in other scenarios, for instance, those using a digital TV set. By stating that the proprietary part of the CAS is exported to the smart card, we unfortunately simplify the model. Although the scrambling algorithm is common for every DVB-compliant system, the STB device hosts some proprietary elements of the CAS. In the early days of Pay TV, the communication between the smart card and the host was in the clear, exposing weaknesses. The control words could be sniffed during transit. For more details on this vulnerability, see "The Devil's in the Details" dedicated to the jugular hack in Sect. 11.2.2. Every CAS provider uses a proprietary protocol that protects the communication between the host STB and the smart card. Thus, the manufacturer of the STB has to implement this proprietary interface. Furthermore, inside the STB, each CAS filters out the EMMs to only transfer the relevant ones to the smart card. This part is also fully proprietary. Were the manufacturer to decide to design a STB that should support several CASs, it would have to implement several proprietary protocols in the same STB, thus increasing its cost. Furthermore, CAS technology providers were worried that an exploit that would break one CAS in the STB may help break the others. Physically separating the proprietary parts would reduce the risk for the other CASs. The DVB consortium decided to find a suitable solution.

The purpose of the DVB Common Interface (DVB-CI) was to export all the elements, including the proprietary ones, participating in Pay TV to a separable component. DVB-CI pushed one-step further the logic that initially exported the management of ECMs and EMMs to the smart card by exporting the entire CAS to a detachable secure module called the DVB-CI module.

Figure 6.8 describes the basic concept of DVB-CI. The STB receives, through its tuner and demodulator,[8] the scrambled transport stream and forwards it to the detachable DVB-CI module. The DVB-CI module demultiplexes the transport stream, i.e., it extracts the different elementary streams such as video, audio, ECM, and EMM. It forwards the ECMs and EMMs to the CAS module. As described in Sect. 6.4, the CAS module returns the control word if the user is authorized to watch the program. The transport stream is then descrambled, and the clear transport stream is passed back to the STB. The STB now acts as if receiving a clear program stream, i.e., it demultiplexes the returned clear stream, decodes the MPEG2 data, and renders it. With this new architecture, every part related to Pay TV is ported into the detachable DVB-CI module. The DVB STB becomes generic and fully independent of CAS flavors. Furthermore, it can easily support several

[8] A demodulator is an electronic circuit that extracts information from the modulated carrier.

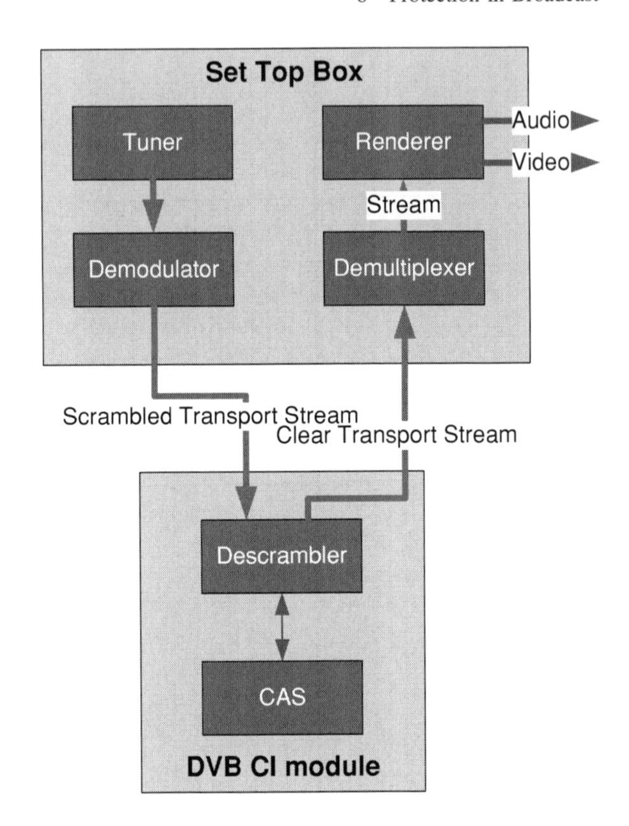

Fig. 6.8 Architecture of DVB-CI

CASs by implementing several DVB-CI slots. In this case, the transport stream is sequentially forwarded to the different DVB-CI modules as a daisy chain.

The DVB-CI physical interface uses a physical connector and a form factor identical to that of the Personal Computer Memory Card International Association (PCMCIA) interface. PCMCIA was the format of an add-on card that was deployed on laptops in the early 2000 s. This format is now obsolete and has been replaced by flash memory cards and the Universal Serial Bus (USB). The DVB-CI module can either embed the CAS module or use an external smart card using the standard ISO 7816 interface. In 1996, CENELEC[9] normalized DVB-CI for the European market [221]. Since 2002, in Europe, DVB-CI has been mandatory for every integrated digital TV (iDTV) set with a screen larger than 30 cm [222].

Figure 6.9 shows a DVB-CI module dedicated to the French digital terrestrial TV. The French digital terrestrial TV, called Télé Numérique Terrestre (TNT), offers 18 FTA channels and several tens of Pay TV channels. A TNT receiver equipped with a DVB-CI interface, for a small cost, can receive Pay TV channels.

[9] CENELEC is the European Committee for Electrotechnical Standardization and is responsible for standardization in the electrotechnical engineering field.

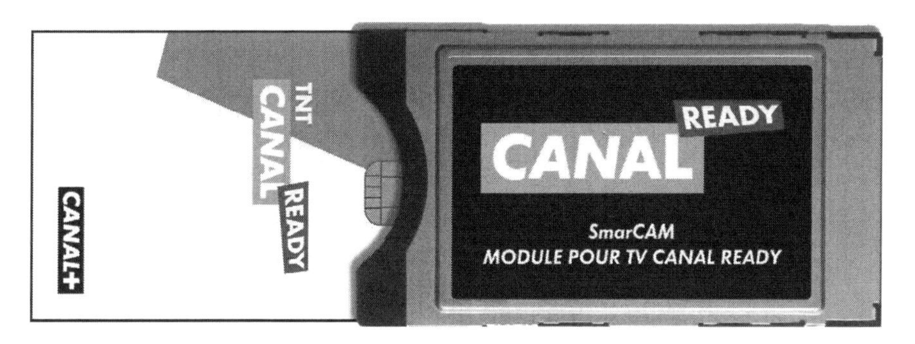

Fig. 6.9 An example of DVB-CI module (© Canal+)

When subscribing to the Pay TV operator services, the consumer receives the corresponding DVB-CI module and the smart card.

The DVB-CI module is not restricted to conditional access. It can be used, for example, for external hard drive or for content processing module. Nevertheless, this possibility has been rarely used in the market.

The reader may have spotted a serious weakness in the DVB-CI architecture. A physical, easily accessible return path that carries digital compressed video in the clear is an obvious hole in the security of the system. Why did the designers not spot such vulnerability? We must remember the context. DVB-CI was designed before 1996. At this time, the main objective was to prevent people from watching Pay TV programs without a legitimate smart card. Nobody was concerned with copy protection in the home. Thus, in 1996, a return path in the clear was not considered vulnerable. A few years later, some network and broadcast operators raised the issue at DVB meetings. This flaw worried content providers and thus they favored integrated solutions for distributing their high-value content. Fixing this problem was a major issue.

In 1998, DVB mandated DVB-CPT to solve the problem. DVB-CPT, focused on home network protection (Sect. 9.5, DVB-CPCM), did not tackle the problem; thus, DVB launched a working group, DVB-CIT, with the mission to address the deficiencies of DVB-CI with a new solution: DVB-CI2. DVB-CI2 had to have a scrambling scheme between the host and the DVB-CI2 module as well as mutual authentication. In September 2007, following some disagreements within the DVB working group, six companies (Panasonic, Philips, Neotion, Samsung, Smart DTV, and Sony) launched a private initiative called CI+. In May 2008, they published CI Plus version 1.0 [223]. In November 2008, they launched a trust authority named CI Plus LPP which licensed the technology and defined the C&R regime.

On top of DVB-CI, CI+ adds a Content Control (CC) system for secure delivery to the host of the content that has been processed by the CAS module.

CC is a complex system. It encompasses many more elements than just content encryption. Figure 6.10 represents the different steps of this protection.

Fig. 6.10 CI+ processing

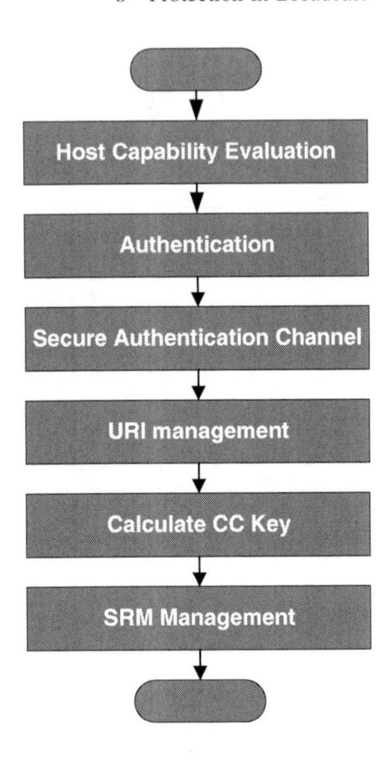

As it is representative of a typical class of protocols, we will spend some time detailing it.[10]

Hosts capability evaluations: As a CI+ module has to be compliant with DVB-CI, it is important for both the host and the module to discover if they are compliant with CI+ or just DVB-CI. The returned content, which was previously descrambled by the CAS module if the consumer was entitled, would only be rescrambled if both the host and the module were CI+ compliant.

All CI+ compliant principals, host and module, have received a 2,048-bit Diffie–Hellman public private key pair signed by the certification authority managing CI+, i.e., CI plus LLCP (Sect. 3.2.3). The signed certificate may be a chained certificate. The capability evaluation is done by mutually checking the signature of each principal's public certificate. If the checking principal validates the chain of the certificate's signatures of the checked principal as OK, it concludes that the checked principal is also a CI+ entity. Both entities will act as defined by Table 6.2. If the verification of the signatures fails, then the CI+ module will not descramble the CAS-protected transport stream even if the customer is otherwise authorized to access the program by the CAS. It will return the scrambled transport stream rather than the clear transport stream. This potentially allows chaining several modules so that a subsequent module may descramble the transport stream.

[10] The explanations will be simplified without going down to the detailed protocols and algorithms.

Table 6.2 CI+ behavior

	Host DVB-CI only	Host CI+
Module DVB-CI only	Act as DVB-CI The returned transport stream is in the clear if the CAS authorized the descrambling	The host emulates a DVB-CI host The returned transport stream is in the clear if the CAS authorized the descrambling The content provider may decide for some high-value content to deny descrambling because it refuses clear transfer
Module CI+	The module emulates a DVB-CI module The returned transport stream is in the clear if the CAS authorized the descrambling Here also, the CAS may refuse to descramble to prevent clear transfer	Host and module act as CI+ entities The returned transport stream is rescrambled if the CAS authorized the descrambling

Authentication: The host and the CI+ module will now mutually authenticate each other. For that purpose, they will prove that they know the private key associated with the certificate they presented each other. Indeed, the single presentation of the certificate does not prove that the verified entity has the private key. The entity could use a stolen or forged public key without actually knowing the companion private key. Therefore, the verification uses a typical challenge-response protocol (Sect. 3.2.3).

As for the host capability evaluation, if the mutual authentication fails, then the CI+ module will not descramble the CAS-protected transport stream.

Secure Authenticated Channel (SAC): The host and the CI+ module establish a SAC. In fact, they define a common session key that will be used to encrypt all exchanged messages. An eavesdropper cannot retrieve this common session key.

URI management: Both sides attempt to find a common Usage Rights Information (URI). The host sends a mask that defines the URI features that it supports. A URI carries the following information:

- `aps_copy_control_info` indicates if the Macrovision analog copy protection has to be activated and with which parameters.
- `emi_copy_control_info` is the equivalent of DTCP CCI. Thus, it supports four modes (copy free, copy once, copy no more, and copy never).[11]
- `ict_copy_control_info`, called Image Constraint Trigger bit, is a Boolean flag that indicates if the output of HD content has to be constrained on the analog outputs.

[11] CI+ uses a different vocabulary than this one, but the semantics is identical. For the sake of clarity, we use the usual CCI vocabulary.

- `rct_copy_control_info`, called Encryption Plus Non-assert bit, is a Boolean flag that indicates if the content can be redistributed or not. This flag is meaningful only for "copy free" content.
- `rl_copy_control_info` defines the retention limit. It is the authorized duration of retention for a record or time shift. It ranges from the default value of 90 min to several days.

The CI+ module receives the URI information from the embedded CAS module and forwards it securely to the host. The host has the responsibility of enforcing the provided URI.

CC key calculation generates the key that will be used to scramble the transport stream.

SRM management is the phase where the CI+ module transfers to the host the potential revocation lists for HDCP and DTCP. These revocation lists, called System Renewability Messages, are provided by the embedded CAS module. Furthermore, the CI+ specifications handle their own revocation mechanism. It is possible to revoke a CI+ module or a CI+ host. For the host, the CI+ module may receive a CRL that defines the set of revoked hosts or a Certification White List that defines the set of revoked hosts that are to be reactivated. DVB-CI does not support revocation. The granularity of these lists is

- a unique host
- a range of hosts
- a certain model of hosts
- a certain brand[12]

Once all these phases have successfully completed, the CI+ module can scramble the transport stream using the CC key. CI+ supports two scrambling methods:

- DES in ECB (Electronic Code Book) mode; this mode is mandatory for every CI+ compliant device.
- AES 128-bit in CBC mode; this mode is mandatory for the HD host.

Like DVB-CI, CI+ supports a way to host in the CI+ module applications that are not CAS-related, such as browsers.

Since its commercial launch in January 2009, about 50 million device certificates have been shipped. The current version of CI+ is V1.3.

[12] The possibility of revoking in one unique CRL a complete set of hosts belonging to one model or even one brand name was too dangerous. If ever an attacker was able to forge such a CRL, then she could create a lethal class attack. The same scenario could occur with a human error. Therefore, the designers of CI+ had to add the Certification White List to possibly counter such class attacks. The usual answer to such an attack is to avoid the granularity of models and brands.

6.6 OpenCable CableCARD

In 1988, the US cable operators established the CableLabs consortium. They launched the OpenCable initiative, which specifies the standards used by the software and hardware of host devices of the cable television industry of the US.

OpenCable protection uses the traditional scheme of Pay TV as discussed in Sect. 6.3. The essence is scrambled using DES in CBC mode. Following the example of DVB-CI, OpenCable defines a standard for a detachable module: CableCARD. It uses the same PCMCIA form factor as does the DVB-CI module. CableCARD has two roles:

- Handle the security to watch protected channels, as done in DVB-CI.
- Handle the navigation among clear digital and analog channels without the need of a STB. It may also support recording. This goal does not exist in the DVB-CI standard. In this book, we will not discuss this navigation feature.

The first version, CableCARD V1.0, is often referred to as the Point Of Deployment (POD) module. The ANSI/SCTE 28 2004 standard [224] defines the interface between the POD and the host. After a ruling of the FCC, all cable providers had to support POD from July 2004. The deployment of PODs was successful. Motorola and Scientific Atlanta are the only suppliers of these modules [225].

The POD uses typical key management based on ECM and EMM. In contrast DVB, the transmission of ECMs and EMMs is done out-of-band (OOB), i.e., it uses a carrier frequency (of about 125 MHz) different from the one used by the MPEG2 signal. A dedicated OOB receiver extracts these ECMs and EMMs before forwarding them to the POD, which exploits them to possibly descramble the protected channels (Fig. 6.11).

Unfortunately, CableCARD V1.0 had several limitations.

- It could only descramble one channel at a time.
- It did not support VOD, PPV, digital program guides, and interactive services. The POD, which was expected to replace STB, offered fewer services than prevalent STBs, and were thus less attractive.
- Although CableCARD V1.0 specifications supported two-way communication, the FCC only mandated one-way receivers, which manufacturers strictly implemented. Thus, the POD could not use all ways of two-way communication inherent in cable networks.

Version 2.0 of CableCARD deals with some of the first version's limitations. It specifies both the receiver and the card as two-way devices. The CableCARD 2 module can simultaneously descramble several streams. To maintain backward compatibility, CableCARD V2.0 defines two modes:

- M-CARD, which supports multiple stream descrambling.
- S-CARD, which only descrambles one stream at a time; this mode is to be compatible with deployed devices supporting only POD.

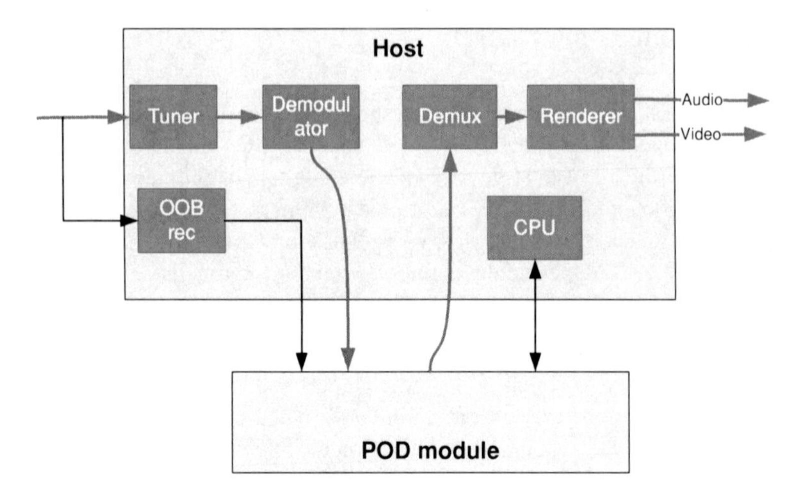

Fig. 6.11 CableCARD V1.0

In contrast to DVB-CI, from the initial design, CableCARD has protected the descrambled program when returned to the host. The scrambling algorithm is DES in the case of S-CARD and triple DES when using the M-CARD mode. The module and the host use a proprietary algorithm, DFAST, to define a common DES key for the scrambling without actually exchanging this key. CableLabs licenses this algorithm as C source code. DFAST is a perfect example of "hook IP" (Sect. 3.8, Compliance and Robustness Rules). With the corresponding patent [226], CableLabs can force licensees to sign the compliance and robustness rules.

The future of CableCARD is darkening. Not many cards of the second version were sold. The Downloadable Conditional Access System (DCAS), an emerging standard, seems to be gaining momentum. DCAS will download an executable software to the host, which will run it. The host will not anymore need a detachable module. For that purpose, CableLabs currently defines a set of Application Programming Interfaces (API) that a compliant host should support. The specification called Open Cable Application Platform (OCAP) [227]. In the case of breach, every OCAP STB can be upgraded by just downloading a new version of the CAS software.

Chapter 7
Protection in Unicast/Multicast

7.1 DRM

DRM usually protects content delivered over broadband connections using unicast and multicast modes. Although functionally very similar to CAS, DRM systems have two significant differences:

- As DRMs have to run on generic computers, they operate in a less trusted environment than does CAS. Their trust model cannot rely on tamper-resistant hardware. Instead, the trust model must rely on software tamper-resistance and easy renewability by forcing the customer to download a new version of the software in the event of a hack.
- DRM assumes the presence (even temporary) of two-way communication, which enables more flexible license management, and some security controls by the remote server.

There are many commercial solutions of DRM, all using similar architecture. As described in Sect. 4.4, Architectural Model, they all have at least three common elements: a content packager, a license server, and a client DRM. The main difference is in the supported usage models. It is extremely difficult to find public information describing security features.

Thus, it is not possible to describe all the available solutions. We will detail only some of these solutions, either because their dominant market share makes them inevitable, or because of their pedagogical potential. We start with the two main actors in the field: Microsoft DRM (Sect. 7.2) and Apple FairPlay (Sect. 7.3). These two examples introduce the basic architecture, a wide range of usage rights, and also some challenges with the individualization of the client. Later, we present Adobe Flash Access (Sect. 7.4) because its design has a distinctive interesting feature. Because OMA is a standard, public information about it is widely available (Sect. 7.5). OMA's description shows a first implementation of domain management. Marlin is a new contender in the arena of

E. Diehl, *Securing Digital Video*, DOI: 10.1007/978-3-642-17345-5_7,

DRM. As its founders are consumer electronics manufacturers, Marlin may be soon deployed in many consumer electronics devices (Sect. 7.6) and thus become a major player.

7.2 Microsoft DRM

7.2.1 Windows Media DRM

Microsoft Windows Media DRM (WMDRM) is probably the most popular and most widely deployed DRM. This is mainly because it is bundled with Windows Media Player. In other words, it is part of every Windows distribution. The first version of Microsoft Windows Media DRM appeared in 1999. In 2001, Microsoft released version 7 of Windows Rights Manager.[1] Due to serious hacks [228], Microsoft issued version 9 the same year. Since 2004, Microsoft has used version 10. At the time of the writing, the current version is 10.1.2.

Microsoft WMDRM v10 natively supports many compression formats and containers: Windows Media Audio V9 (with the `.wma` extension), Windows Media Video V9 (with the `.wmv` extension), three proprietary versions of MPEG4, and the ISO-defined MPEG4 format.

Microsoft WMDRM has four elements:

- The Packager prepares the piece of content to protect.
- The Windows Media Server stores the packaged piece of content and distributes it to the customer.
- The Windows Media Rights Manager generates the licenses.
- The Windows DRM client handles the delivered piece of content and license. If allowed, it delivers or renders the clear piece of content. The client is often hosted in Windows Media Player. Nevertheless, device manufacturers or system integrators may implement the Windows DRM client in their own media player.

The architecture of WMDRM, illustrated in Fig. 7.1, is classical. The Packager receives clear content. It encrypts the essence using a symmetric encryption algorithm (AES) using a 128-bit random key. The random key is generated from two pieces of information:

- The license key seed is a secret data only known to the Windows Media Rights Manager and the content owner who delivered the clear content. The license key seed is unique for each piece of content to be protected.
- The Key ID, often referred to as the KID, is public information that is unique for each Windows Media file.

---

[1] Indeed, version 7 was the second commercial version.

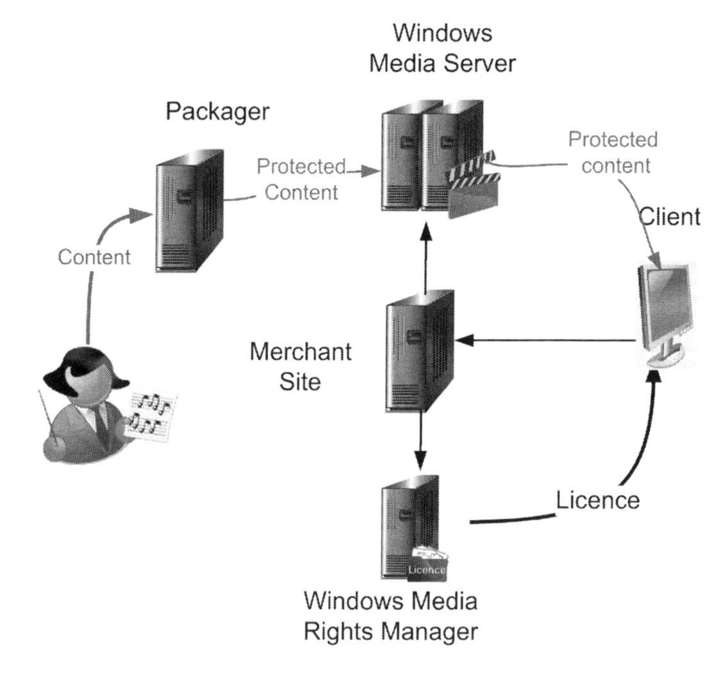

Fig. 7.1 Architecture of Windows Media DRM

The Packager appends a header to the encrypted essence. The header contains at least the following information:

- Key ID.
- Content ID is a unique reference to the actual piece of content.
- Version number, which defines the minimal needed version of DRM client; the version number ensures that the content will be played back with the expected level of security. A DRM client whose version number is smaller than the requested one should not grant access to the piece of content. The latest pieces of content should preferably use the latest version number, i.e., the most secure one.
- Individualized version attributes.
- License acquisition Universal Resource Locator (URL) is the address of the license server that can deliver a valid license for this piece of content.

The header is in the clear. Nevertheless, a signature protects its integrity.

The Devil's in the Details
Why should the header be protected in integrity? A superficial analysis may give the false idea that if ever a cracker modifies any parameter of the header, it would spoil the descrambling. For instance, changing the key ID would only result in a wrong decryption key that would not descramble the

essence. Changing the content ID would have the same result: the DRM client would query a license for a different piece of content, thus receive a wrong license, and fail to descramble the content. Similarly, changing the version number, for instance, for a smaller version number, requesting a weaker implementation, would be useless because usually the license format differs from one version to another. Thus, we could conclude that tampering with the header is useless.

Unfortunately, changing the URL has interest to crackers. A pirate may create a malicious file. The cracker would take a genuine WMDRM-protected file. She would replace the license acquisition URL with the address of her own malicious site. She would post the forged file on sharing sites such as P2P networks. When a WMDRM client attempted to access this file, it would look for a license that it did not yet have. Then, it would connect to the license server defined by the forged license acquisition URL. In other words, it would connect to the malicious site. This site may, for instance, attempt to infect the connecting computer. It could be a stealthy operation. Indeed, WMDRM could be turned into a kind of contamination vector.

If an authority signs the header, it is not anymore possible for the cracker to forge another license acquisition URL than the URL of its legitimate license server. At least two Trojan malwares exploited this attack with WMDRM v10: WmvDownloader-A and WmvDownloader-B [229]. A countermeasure has been implemented. The header is now signed; hence, the URL cannot be anymore tampered with.

Lesson: The attacker may derail your system for another purpose than the one you expected. Protect the integrity of your information.

Figure 7.1 describes the transactional model of WMDRM. The consumer purchases a piece of content, for instance, a song, from a merchant site. Once the commercial transaction has completed, the merchant server requests the Windows Media Server holding the corresponding protected item to deliver the purchased protected piece of content to the consumer. The delivery may be file-based or stream-based. To play back the received piece of content, the consumer's DRM client needs the corresponding license. Each DRM client has on its host a protected license store. The DRM client checks first if the license is already available in this license store. If this is not the case, or if the license is not anymore valid, the DRM client requests a valid license from the Windows Media Rights Manager operated by the merchant. For that purpose, it connects to the Windows Media Rights Manager at the URL defined by the license acquisition URL in the header of the piece of content. The DRM client transmits also information from the content header (such as the Key ID) and platform information that uniquely describes the system on which the DRM client executes. Among this platform information is the public key of the DRM client. For that purpose, WMDRM v10 uses 160-bit Elliptic Curve Cryptography (ECC).

Every consumer receives the same installation package. Therefore, once installed, all DRM clients are identical and do not hold an individualized, unique key pair. The DRM client acquires its public private key pair during the personalization phase, which occurs when the customer attempts to access her first piece of protected content. In fact, a generic DLL[2] (with a nonstandard extension[3]) is transformed into an individualized DLL that contains the private key in an obfuscated form. In other words, the 160-bit ECC private key is spread all over the software of the DLL. It is thus extremely difficult to retrieve the private key (Sect. 3.6, Software Tamper-Resistance). The individualized DLL is linked to the host computer, i.e., the same instance of this DLL would not work on another host computer. Of course, the DLL holds also the public key. The public key does not need to be protected. A remote server drives this personalization process.

The Windows Media Rights Manager analyses the received set of information. It identifies the content owner and checks the rights associated with the requested license. Once it has verified that the requester fulfils the mandatory conditions, the Windows Media Rights Manager calculates the random encryption key using the secret license key seed, which it already knows, and the Key ID, which it receives from the DRM client. Then it builds the corresponding license that holds this encryption key and the usage rights granted to the customer. Using the platform information, the Windows Media Rights Manager cryptographically binds this license to the targeted DRM client. In fact, it encrypts the license with the 160-bit ECC public key of the DRM client.

When receiving the license, the DRM client stores it in its license store. The DRM client is now ready to descramble the piece of content.

In fact, WMDRM v10 supports three modes of license delivery:

- *Silent license delivery* directly sends the license to the DRM client. This assumes, of course, that the consumer already paid for her purchase.
- *Non-silent license delivery* redirects the consumer to a webpage. The webpage may be used for registration or payment. The Windows Media Rights Manager delivers the license only once the expected task is successfully completed on this webpage. In other words, the Windows Media Rights Manager waits for an acknowledgement prior to issuing the proper license.
- *License pre-delivery* is similar to silent license delivery except that the license is sent to the DRM client prior to the DRM client's request, i.e., before the DRM client actually needs the license. This may enhance the user experience, for instance, when using subscription-based models. If the license pre-delivery occurs early enough, there is no risk of interruption of services when renewing

---

[2] A Dynamic Link Library (DLL) is a collection of functions that is invoked as an external library by a main program. A DLL is loaded at run time when one of its functions is required. This is different from static libraries that are linked after compilation. Static libraries are loaded with the main program.

[3] In Windows OS, `.dll` is the standard extension for a DLL.

the subscription period. The pre-delivery date of the license is at the initiative of the content owner.

WMDRM supports many usage rights such as the following:

- *Time-limited playback*: this type of usage rights is only available for devices that have a secure clock, or networked devices. In the latter case, the internal clock synchronizes with a web-secure clock service. The license may contain the following information:
  - *Begin Date*: Before this date, the consumer cannot access the piece of content even if the protected content and the license are already available on the computer.
  - *Delete on Clock Roll Back*: The DRM client destroys the license if ever the current time of the hosting computer has jumped to the past. This prevents trivial attacks where the attacker modifies the current system time to bypass expiration of the license[4] so that the system time never reaches the expiration date. To access again the corresponding piece of content, the DRM client will have to request a new license. This request may allow the DRM client to resynchronize its internal clock.
  - *Disable on Clock Roll Back*: The license is disabled if ever the current time has jumped to the past. Reversing the roll back, i.e., coming back to a time later than the last memorized value, will reactivate the license without the need for a query to the Windows Media Rights Manager. This mode is less "violent" than Delete on Clock Roll Back.
  - *Expiration after first use*: This defines how long the consumer can access the piece of content after having accessed it once. The expiration date is a relative value.
  - *Expiration date*: The piece of content cannot anymore be accessed after this date regardless of when it was first watched or listened to. The expiration date is an absolute value.
  - *Expiration on Store*: This defines how long the license is valid after the acquisition of the license from the server. The expiration date is relative to the acquisition time, regardless of the actual use of the corresponding piece of content. In other words, the consumer loses his right to access the piece of content even if he never watched or listened to it.
  - *Grace Period*: This defines an extension period after the expiration period defined by the previous parameters. This option is extremely useful in the case of subscription-based models, for preventing a bad user experience if the license is not renewed or delivered on time.

[4] Obviously, the detection of clock roll back takes into account the automatic time shift due to summer time adjustments.

- *Limited number of playbacks*: The DRM client counts the number of playback operations. If the counter exceeds the value defined in the license, then the DRM client requests a new license from the Windows Media Rights Manager. If the license is not renewed, then the DRM client does not grant access to the piece of content.
- *Transfer to a portable device*: The Windows Media Rights Manager may generate a new license bound to the portable device to which the content is exported. The Windows Media Rights Manager may count the number of copies and possibly limit the number of generated licenses.

WMDRM v10 for portable devices introduces an interesting new concept: *chained licenses*. The chained licenses are part of a classical hierarchical tree. The "leaf" licenses correspond to usage rights for a given piece of content. The "root" license is more generic and applies to all the leaves. In other words, the leaf license inherits validity from its parent license. For instance, in the case of a subscription model, the root license may manage the end date of the subscription period through the expiration date parameter. The leaf licenses grant basic access to each concerned piece of content. Thus, as long as the root license is valid, all pieces of content acquired by the user are valid. At the end of every subscription period, the merchant issues a new root license valid for the coming subscription period. There is no need to individually renew every single license for each acquired piece of content. Leaf licenses may have additional, more restrictive usage rights than the root license.

Like all other DRMs, WMDRM has been successfully hacked several times. In 1999, "`Unfuck.exe`" was the first WMDRM hacking software. The crackers found a way to capture a compressed song just after its descrambling and before it was transferred to the audio decoder. The result had better quality than that using the analog hole (Sect. 9.6, The Analog Hole) because it created a perfect digital copy without needing any compression. It operated in the compressed domain. Obviously, the hack required a legal license to play back, at least once, the protected audio file. In 2001, a cracker by the name of "Beale Screamer" issued a new hack "`Freeme.exe`" together with a complete, technical, detailed description of MWDRM version v7 [230]. `Freeme` used a legal license to retrieve a full version of any content. In fact, Beale Screamer discovered the way to decrypt the license using one weakness. With a decrypted license, `Freeme` was able to descramble the content. `Freeme` did not exploit the analog hole. It was the result of careful reverse engineering of the DRM client. In 2006, a cracker by the name of Viodentia released the first version of "`Fair4Use`". The latest version of this hack is "`FairUse4WM 1.3`" and only works for Windows XP and a given version of WMDRM v10. When WMDRM v10 handles a protected key, it stores the key in the computer's volatile memory. `FairUse4WM` extracts this information. Very quickly, Microsoft issued a patch that corrected the weakness. Interestingly, `FairUse4WM` uses tamper-resistance techniques (Sect. 3.6, Software Tamper-Resistance) to protect itself [134].

At the time of writing, no hacking tools are available for the version of Windows DRM used by Windows Media 11.

7.2.2 PlayReady

Since 2008, Microsoft has relaxed the constraints for implementation with its new DRM system, PlayReady [231], with two main evolutions:

- PlayReady is content agnostic, i.e., it scrambles files or streams independently of their format. Thus, PlayReady can protect any type of digital content including video, audio, games, wallpapers, ringtones, and text documents.
- PlayReady supports other operating systems than Windows. In addition to the different flavors of Windows, PlayReady is also compliant with Mac OS using Silverlight. Silverlight is Microsoft's browser plug-in for rich media [232]. It is cross-platform and cross-browser. PlayReady can also be ported to Linux for embedded devices. Nevertheless, Microsoft does not plan to support Linux for computers (even with Moonlight, the Linux port of Silverlight [233]).

As illustrated in Fig. 7.2, PlayReady uses several components:

- The content packaging server that prepares the content to be protected and also the domain certificates.
- The distribution server that holds the protected content and delivers it to the client. It can be a normal Web server.
- The license server that generates the licenses. A license can be bound to a device or to a domain.
- The domain controller that manages the registration and credentials of devices belonging to a domain.
- The metering server that collects anonymous usage information for accurately sharing the revenues.
- The client that renders the protected content if authorized.

We will highlight the main differences with Windows Media DRM. The first evolution is the support of domain. A domain encompasses the set of devices belonging to a customer, a family, or a group of users. Content can be bound to a domain. For a client to access domain-bound content, it has first to join the domain. This is the role of the domain controller, which checks that there are not already too many devices in the domain. Once successfully registered, the client receives the credentials of the specific domain that permit decrypting licenses bound to it. Of course, it is possible for a client to leave a domain. The distributor defines the policy and size of the domain, following commercial agreements with content owners. A client may belong to many domains, depending on the policies of distributors. Distributors can define new domain-based business models, such as home rental or subscription encompassing any device of the household (as long as they are registered in the same domain with the domain controller).

The licenses have evolved so that a PlayReady license can be bound to a domain. Then, every client belonging to the domain can access associated content. The license is encrypted using a key known to all the members of the domain. The domain controller delivers this key during the registration phase. Furthermore, the

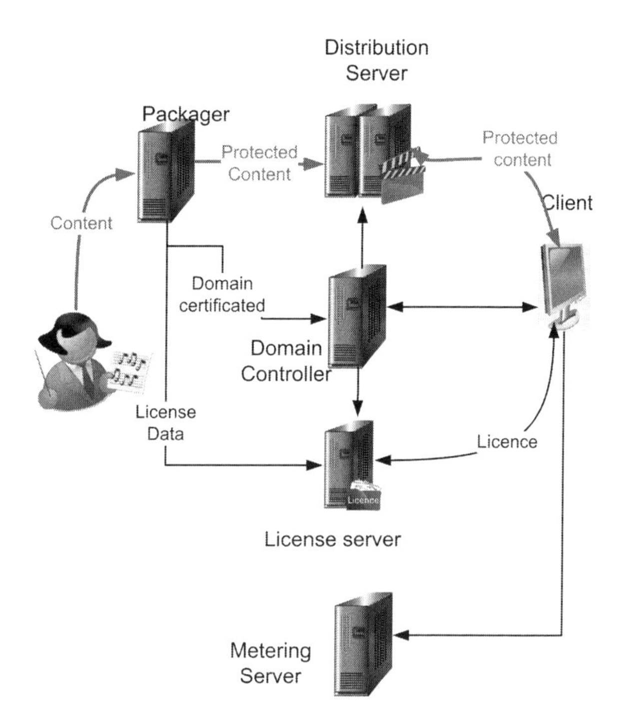

Fig. 7.2 Architecture of PlayReady

license can be embedded in the protected content itself. This simplifies the distribution. If the license server knows all the information related to the target client(s) before distribution, it can provide the license to be embedded.

A PlayReady client can play WMDRM-protected contents. The client understands the WMDRM licenses. WMDRM clients are not compatible with PlayReady, i.e., a WMDRM client cannot access PlayReady content. Furthermore, a PlayReady license server can deliver WMDRM licenses, whereas a Windows Media Rights Manager cannot deliver PlayReady licenses.

PlayReady is compatible with Microsoft's new interoperable format PIFF (Sect. 11.3.2).

7.3 Apple Fairplay

In January 2001, Apple opened the iTunes store to boost the sales of iPods. For iTunes to be attractive to consumers, it had to offer a large catalog of good songs. Apple had to negotiate with the major studios. DRM was about their main requirements. To please them, Apple had to propose a DRM protecting the sold songs. This was the *raison d'être* of FairPlay. Nevertheless, it was always clear that Apple was not in favor of DRM. In a very famous manifesto [234], Steve Jobs predicated (and even called for) the end of DRM.[5] It is important to keep in mind that Apple's main business model is the sale of hardware rather than software. Although iTunes is profitable, the income and profit from the sale of hardware devices such as Pod, iPhone, iTouch, and iPad exceed by far those from iTunes. A wide ecosystem with attractive songs and applications is just an additional incentive to purchase Apple's devices. Furthermore, the use of proprietary formats and DRM is an efficient way to lock consumers to the brand's product lines [235]. The more songs a customer will have purchased on Apple's store, the more expensive it will be for the customer to switch to another brand. She will have to purchase not only a new device but also her complete library of content.

The usage rights supported by FairPlay are simple and build a rather "balanced" set of copy restrictions that may be reasonably acceptable to customers. A FairPlay-protected piece of content has, by default, all the following associated usage rights:

- It can be duplicated on an unlimited number of iPod devices but still be protected by FairPlay.
- It can be duplicated on five different computers and still be protected by FairPlay.
- It can be burnt on an unlimited number of audio CDs without any DRM protection.

Apple uses the mp4 container format, with the audio tracks being encoded as Advanced Audio Coding (AAC). Every piece of content is scrambled with a master key using AES. When a customer purchases one piece of content on the iTunes store, the store registers this purchase on a dedicated server with the identification of the customer. In return, the server provides a random key, called a user key. The store encrypts the master key with the user key. The encrypted master key is embedded in the mp4 container together with the scrambled content. The iTunes server securely delivers the user key to a protected directory (`/Users/Shared/SC Info`) inside the customer's device. The protected directory contains a file, `SC Info.sidb`, that holds the user key of each piece of content purchased by the customer. Both the directory and the file are invisible system files. When playing protected content back, the FairPlay client looks for the user key protecting the corresponding master key in the key database `SC Info.sidb`. If the user key is available, the FairPlay client decrypts the encrypted master key using the user key. With the recovered master key, the FairPlay client descrambles the content. If the corresponding user key is not available in `SC Info.sidb`, the device will connect to the iTunes store (Fig. 7.3).

Of course, one of the most important assumptions of the FairPlay system is that the database of user keys, `SC Info.sidb`, is secure. Several reverse engineering

[5] Initially, iTunes offered FairPlay-protected songs for $0.99 per tune. In 2007, Apple offered also DRM-free songs for $1.29 [28].

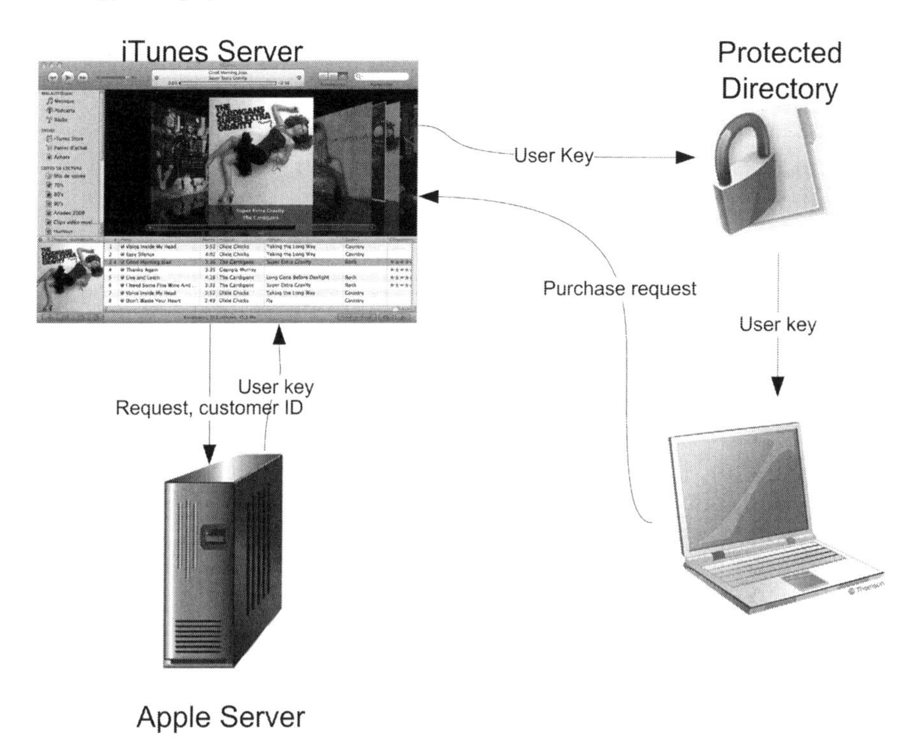

Fig. 7.3 Apple's FairPlay infrastructure

exploits shed some light on the implemented solution. The database is encrypted using a two-step process:

- The plaintext database is XORed with the output of a proprietary Pseudo-random Number Generator (PRNG) seeded with unique information extracted from the computer.
- Then, it is AES encrypted with one more unique piece of information extracted from the system of the computer (Fig. 7.4).

The Devil's in the Details

The encryption scheme of the database of user keys is interesting. Encrypting the database using AES with a key unique to the computer is a typical method of pairing the database with the hosting computer. The key is derived from unique parameters of the computer. This technique is sometimes called hardware soldering or machine fingerprinting. Should the attacker transfer this database to another computer, the decryption will fail because the key derived from the new computer will be different. The basic assumption is that the fingerprint of a machine is unique and non-forgeable. The actual security relies on the methods that build this unique computer key.

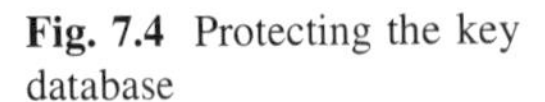

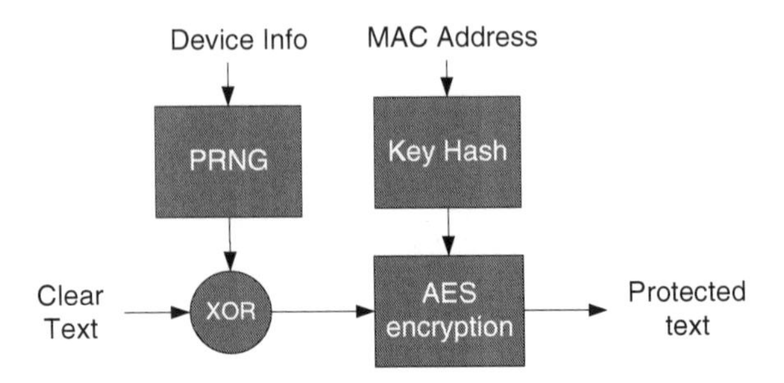

Fig. 7.4 Protecting the key database

The computation uses several physical parameters of the host such as MAC address, serial number, hard drive number, memory size, and configuration of the OS.
In the case of FairPlay, it seems that the computer key is an obfuscated hash of the MAC address for a Mac OS X based client and the result of a hash using registry keys and drive information for a Windows OS based client. Unfortunately, all these parameters are mostly public, and thus fragile. A serious hacker should be able to write low-level software intercepting the system calls and impersonating a given device. In other words, it is not extremely robust against attacks. Secure machine fingerprinting is a promising research field. A new type of solution appears with Physical Unclonable Function (PUF) chips [236]. Microelectronics processes introduce randomness. No two chips, even on the same wafer, are strictly identical. A PUF uses this randomness for uniquely identifying the instance of this chip without the possibility of cloning. PUFs are not yet widely deployed, therefore, currently machine fingerprinting is not extremely secure.
To mitigate this weakness, most probably, Apple engineers added a first encryption that uses the output of a proprietary PRNG. The design assumes that the actual algorithm of this PRNG remains secret. In other words, it is a secret function. Thus, to mitigate the risk of a not so secret machine key, Apple added a secret function. Unfortunately, this secret function did not remain secret for long [237].
Lesson: Without tamper-resistant hardware, it is difficult to keep information secret. Proprietary secret functions can partially replace secret information. Unfortunately, proprietary secret functions are in the category of "security by obscurity". As seen previously, security by obscurity never survives very long.

As described previously, Apple authorizes the protected content to be copied on five different computers. How does this work? A machine ID uniquely identifies each computer. A machine ID is a data computed from different information of the computer and is supposed to be non-forgeable. In other words, it should be impossible to build a computer that would present a given ID. To be authorized to

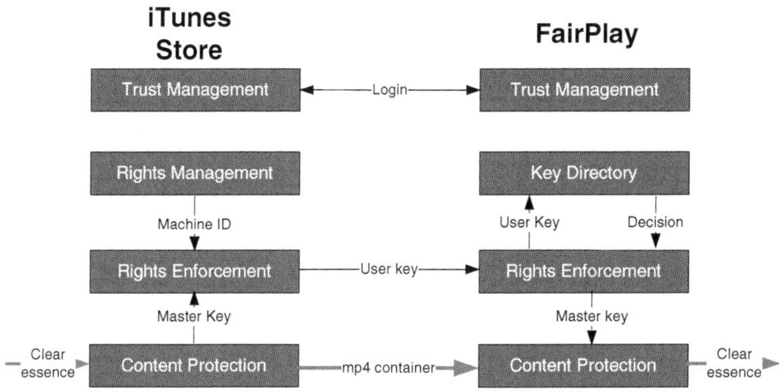

Fig. 7.5 Four-layer model for FairPlay

transfer a piece of content to a computer, this computer must be first registered in a database managed by Apple. Therefore, the FairPlay client sends to the iTunes store its machine ID with the identity of its owner. The machine ID is forwarded to the Apple server, which checks that the user has not already registered five computers. If this is not the case, the server registers the new computer. Then it securely returns all the user keys of the user to the FairPlay client, which stores them in its protected directory. Then the user can access all the pieces of content he purchased and transferred on this new computer.

The transfer to an iPod or iTouch is even simpler. The FairPlay DRM client transfers the user key together with the corresponding piece of content to the protected directory. With the corresponding user key, the iPod can extract the master key and thus descramble the corresponding protected content. It does not need to interact with Apple's server because, by default, content can be transferred to any iPod. In contrast, in the case of transfer to computers, an authority must enforce the threshold of five computers; hence the mandatory interaction with Apple's server.

Figure 7.5 represents FairPlay DRM with the four-layer model. Because FairPlay defines a fixed single common set of usage rights, the rights management is simplified. Mainly, the server maintains for each consumer a list of registered computers and a list of purchased pieces of content. On the client, the user key acts like a license. Having the user key grants it access to the corresponding piece of content. Trust management is mainly ensured by the login to the iTunes store and the scheme of protection of the protected directory `SC Info.sidb`. Furthermore, the proprietary PRNG is a way to assess the compliance of the FairPlay.

The first person to "hack" FairPlay was the (in)famous Jon Johansen (known as DVD Jon; Sect. 8.2, CSS), with `QTFairUse` software. This software records directly from the computer's RAM the digital sound descrambled by FairPlay and delivers a raw AAC file. Other software can then encapsulate this raw AAC data into friendlier containers such as mp3. As such, `QTFairUse` is a sophisticated variant of the "analog hole" (Sect. 9.6, The Analog Hole). Later, DVD Jon released `DeDRMS`, a piece of software that directly descrambles the protected content using its user key. Does it fully break the DRM? Strictly speaking, FairPlay has not yet been broken. The hack needs legally purchased content, i.e., it needs the actual user key. The software removes the encryption but `DeDRMS` does not remove the User ID, name, and email address. Interestingly, Apple has never sued Jon Johansen. `QTFairUse` uses a principle similar to that of `UnFuck`.

Other software such as `JHymn` or `Requiem` have also "hacked" FairPlay. `JHymn` cracks only iTunes 6. Since 2008, successive versions of `Requiem` directly decrypt the user keys stored in the protected database, thus, allowing access to the clear content beyond the rules defined by Apple.

7.4 Adobe Flash Access

In May 2008, Adobe introduced its own version of DRM: Adobe Flash Access. The current version, Adobe Flash Access 2.0, only supports the container formats `.flv` and `.f4v` for Flash video. The client application, running under Adobe AIR[6] or with Adobe Flash player SWF, is available for Microsoft Windows, Apple Mac OS, and Linux.

As illustrated by Fig. 7.6, the Flash Access system uses a rather conventional DRM architecture. Nevertheless, its data flow is different from the usual one.

- The packager prepares the content. Each packager has a unique, certified RSA 1024 public/private key pair signed by Adobe. Indeed, the packager generates its own public/private key pair and requests Adobe to sign its public key. Thus, Adobe never knows the packager's private key.
- The client is integrated with the Flash player or Adobe AIR runtime. The video player SWF or the AIR application provides the user interface for the client. It receives the protected content and, if authorized, displays it. Each client has a local store for its licenses. Each client has a unique key pair issued by Adobe.
- The license server delivers the license to the client. Each license server has a RSA 1024 unique public/private key pair signed by Adobe.

The preparation phase, or packaging, consists of the following steps:

- Encode, if not yet done, the video file in `FLV` or `F4V` format.
- Define the policy, i.e., the default usage rights associated with this piece of content. Later, the license server can overwrite the default policy when generating the actual license. The default policy may be empty. In that case, the license server defines a new policy when generating the license. Flash Access supports a combination of the following constraints [238]:

[6] Adobe AIR is an execution environment that interprets HTML, Java, Flash, and Action Script without the need of a browser.

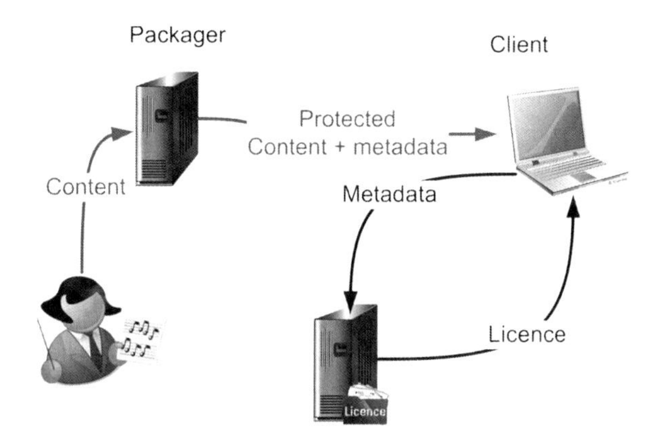

Fig. 7.6 Adobe Flash Access synoptic

- *Start date* defines the date after which the piece of content can be viewed. This parameter is similar to Windows Media DRM's begin date.
- *End date* defines the date after which the piece of content cannot be anymore watched, at least with the current license. This parameter is similar to Windows Media DRM's expiration date.
- *License caching duration* defines how long the license can be cached in the client's local store. During this period, the client does not need to ask the license server for a new license. The duration is defined either in relative time (number of days) or by an absolute date. This parameter is similar to Windows Media DRM's expiration on store.
- *The playback window* defines how long the piece of content can be viewed after its initial playback. This parameter is similar to Windows Media DRM's expiration after the first use.
- A client may be required to authenticate the license server prior to acquiring the license.
- *The list of SWF player/AIR applications* authorized to access the content. It is thus possible to limit the set of applications that can view it. Each authorized application is identified by its application ID, the publisher ID, and a range of valid version numbers. For instance, a content distributor may restrict playback of its content to its own players.
- *The black list of DRM clients* defines the minimal version of Adobe Flash Access and the execution platforms needed to access the content. This allows phasing out compromised versions. The license server checks the validity of the requesting client before delivering the license. Each DRM client is qualified by a security level, a version number, and the supported operating system.
- *Custom usage rules* are non-Adobe data carried with the license. The client forwards them to the custom client code that will apply its own enforcement strategy and thus partly supersede Adobe's usage rights. Nevertheless, it can

only apply more restrictive usage rights. If initially Flash Access does not grant access, then the custom client cannot override this decision.
  - *License chaining* is similar to Microsoft's chained licenses. The parent root license allows the update of a full batch of chained licenses.
  - *Output protection controls* define requirements on the video output. Four levels of requirements separately control the analog and digital outputs: protection required, use protection if available, no protection needed, and do not output.

- Once the default policy has been defined, the packager encrypts the piece of content using a random Content Encryption Key (CEK).[7] Adobe has defined three levels of encryption:
  - High-level encrypts the full H264 data.
  - Medium-level only encrypts about half of the H264 data.
  - Low-level encrypts between 20 and 30 % of the H264 data.
  - The lower levels of security target devices with reduced capacities. For instance, if the descrambling is performed by software implementation rather than hardware implementation, then it may be interesting to reduce the quantity of data to descramble to unburden the processor.

The CEK is stored in the metadata associated with the media file. For that purpose, it is encrypted using the public key of the targeted license server. Thus, only the license server associated with this public key can access the encrypted CEK. A URL defines the location of the corresponding license server. The metadata are signed with the private key of the packager to ensure integrity of the data. Furthermore, this signature ensures that only a trusted packager can create the metadata. A rogue packager cannot properly sign the metadata.

The protected piece of content, together with the metadata, can now be delivered to the distributors or directly to consumers. It is interesting to note the specificity of Adobe's design: there is no direct connection between the packager and the license server. The only link is the certified public key of the license server. The Flash Access license server does not have persistent knowledge of the CEK used to protect the pieces of content. To retrieve the CEK, the license needs the metadata of the protected content. This approach is different from the usual one.

The playback of protected content consists of the following steps:

- If the client does not yet have a valid license, it requests it from the license server by sending its credentials and the metadata of the protected content. A URL in the metadata defines the address of the license server. The client may

[7] Indeed, the packager may define a given value for the CEK rather than a random one. For instance, the packager may use the same key for a period of 24 h. This is useful for use cases such as linear content, i.e., without a beginning or end. Another use case is the use of the same CEK for different encoded instances (HD, SD, mobile) of the same piece of content.

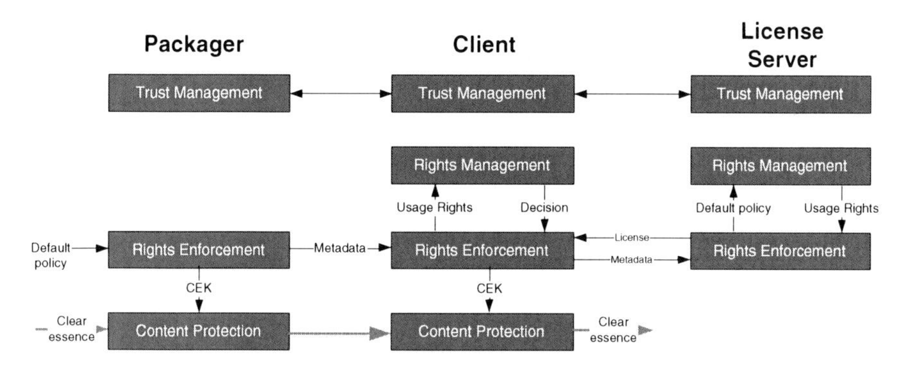

Fig. 7.7 Four-layer model of Adobe Flash Access

have a valid license cached in its persistent license store, in which case it descrambles the protected content without having to ask the license server.

- Once the signature of the metadata has been validated (using the public key of the packager), the license server examines the default policy that was part of the transferred metadata. The license server checks in its database that no policy overwrites the default policy defined by the metadata.
  If the packager always defines a default policy that should be the final license, then it is possible to use an extremely simplified stateless license server with minimal information.
- If the policy requires authentication, then the client will have to present valid credentials. The application may query a user login/password or collect a cookie. The license server may request the credentials or receive them in advance in the metadata delivered by the application.
- If the license server decides to grant access, it decrypts (using its own private key) the encrypted CEK. It can now build a license containing the CEK and usage rights.
- The license server encrypts this license with the client's public key and delivers it to the client.
- The client decrypts the license. After having checked its compliance with the usage rights in the license, the client descrambles the protected content with the CEK.

As usual, before being able to use Flash Access DRM, the client has to be individualized. The client application connects to an Adobe individualization service, which provides a key pair unique to this instance of the client. This key pair cryptographically bounds the license to the client.

Figure 7.7 illustrates the four-layer model of Flash Access. The model illustrates the specificity of Flash Access, i.e., the absence of a direct link between the packager and the license server. The client serves as an intermediary for the transfer of the CEK to the license server. The content protection layer is only managed by the packager and the client.

7.5 Open Mobile Alliance

The Open Mobile Alliance (OMA) was formed in June 2002 by nearly 200 organizations including the world's leading mobile operators, mobile phone manufacturers, network suppliers, information technology companies, and content and service providers. OMA develops specifications to support the creation of interoperable end-to-end mobile services [239]. Among these services, OMA defines a DRM to protect audiovisual content sent to mobile phones. OMA issued a first version, so called OMA 1.0, whose target was mainly to protect ring tones. The specifications defined three delivery modes:

- Forward lock: The piece of content is sent in a file with extension .dm through either file download or Multimedia Message System (MMS). The piece of content is not encrypted. The only constraint is that the receiving device is not allowed to forward the file to another device. The piece of content can be accessed without any restriction.
- Combined delivery: The essence and the rights object, and the terms of the license in OMA's lexicon are sent together in a file using the same extension, `.dm`. The file can be directly downloaded or sent by MMS. Here also, the content is not encrypted and cannot be forwarded. The rights object defines the number of times or the period for which the content can be accessed.
- Separate delivery: The essence is encrypted in a format called the DRM content format (with a symmetric algorithm) and transferred in a file using file extension `.dcf`. The transfer can use any channel such Bluetooth, IrDA, file download, MMS, or even email. The rights object is packaged with the key used to encrypt the content in a message using file extension `.drc` or `.dr`. The supported usage rights are identical to those of combined delivery. The rights object is delivered using a secure channel. The receiving terminal may forward the encrypted essence to another device but not the rights object. In other words, separate delivery enables super distribution.

OMA DRM 1.0 assumes that the terminal is reliable and that the communication channels are secure. In forward lock and combined deliveries, the essence is in the clear. In separate delivery mode, the encryption key is in the clear within the `.drc` message. With the promise of downloadable movies, or songs, arose a need for a more secure DRM to protect such media content. The former assumptions were not anymore acceptable to deliver valuable contents. Thus, OMA designed a second DRM version: OMA DRM 2.0 [240].

OMA DRM 2.0 uses a very traditional architecture, as illustrated by Fig. 7.8 with three actors:

- content issuer that handles the content;
- rights issuer that generates the license; and
- DRM agent that is the DRM client executing on the mobile appliance.

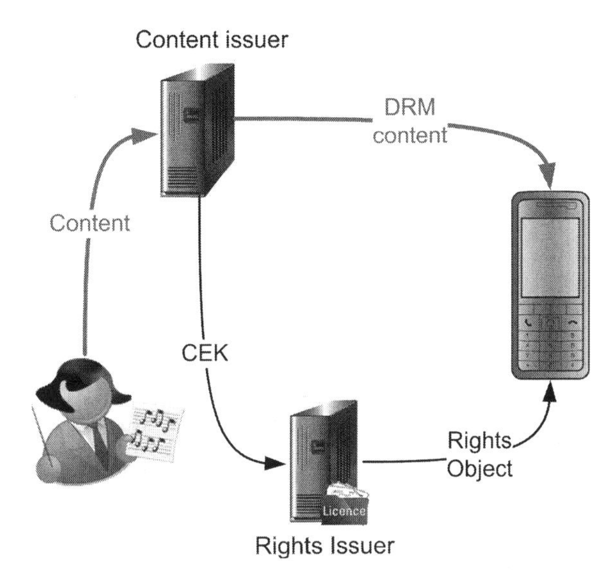

Fig. 7.8 OMA architecture

A content issuer packages media objects into DRM content and protects them. It scrambles clear media objects using 128-bit AES with a symmetric Content Encryption Key (CEK). The result is an encrypted content file with file extension `.dcf`. Although not mandatory, we may expect the CEK to be different for each piece of content.

The rights issuer defines the associated usage rights. These usage rights are described using ODRL as REL (Sect. 3.7, Rights Expression Language). These usage rights are packaged together with the CEK into a rights object. The rights object is cryptographically bound to the DRM content it specifies. Normally, a rights object is associated with one DRM agent. Every DRM agent has a unique public/private key pair with a certificate delivered by a CA. In addition to typical PKI properties, the certificate carries also information on the characteristics of the DRM agent. Thus, a rights issuer may decide whether to deliver rights object to a given DRM agent or not. The rights issuer encrypts critical parts of the rights object with the public key of the DRM agent.[8] Then, the rights issuer computes a cryptographic hash of the rights object and signs it with its own private key (Fig. 7.9).

OMA 2.0 supports the new concept of domain. A domain defines a set of portable devices. A rights object may be associated with all the DRM agents belonging to the same domain. The definition of the policy managing the membership in the domain is under the responsibility of the rights issuer, or, more precisely, of the content distributor.

[8] Indeed, OMA V2 uses two layers of encryption. Within the rights object, the CEK is encrypted with AES-WRAP using a Rights Encryption Key (REK). Only then is the rights object encrypted with asymmetric cryptography.

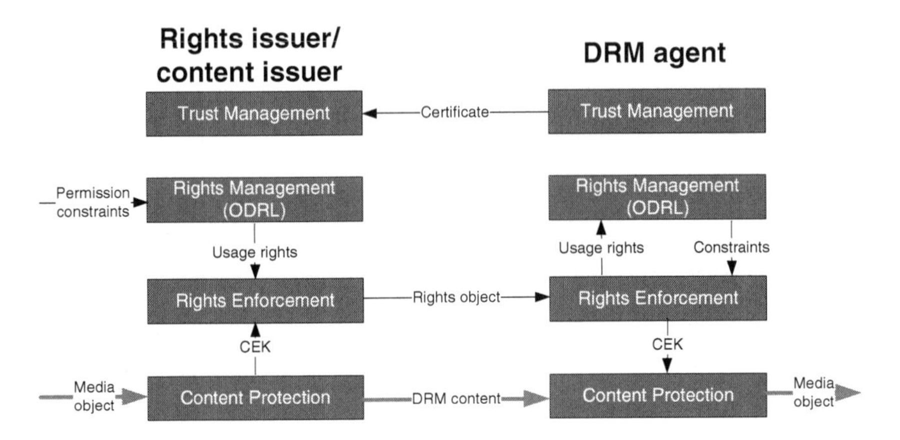

Fig. 7.9 Four-layer model of OMA DRM 2.0

Content issuer and rights issuer are roles. Thus, they may be operated by the same entity or distinct entities.

A DRM agent is the trusted entity that executes on the mobile appliance. The DRM agent receives the DRM content and the rights object. The DRM content (`.dcf` file) and rights object may be delivered together or separately at different times, depending on the selected delivery mode. When trying to read the `dcf` file, the DRM agent opens the associated rights object. This would only be possible if the rights object was built for this instance of the DRM agent or for a domain that the DRM agent joined. The rights object is encrypted so that only the targeted DRM agent can decrypt it. The rights management layer parses the ODRL expression to check the permissions and extracts the constraints. The rights enforcement layer ensures the obedience to these constraints and passes the CEK to the content protection layer, which descrambles the DRM content if the rights enforcement granted it access (Fig. 7.10).

In addition to defining the protection of content, OMA also defines the way DRM objects and rights object may be delivered to devices. OMA supports pull mode via http or PMA download, pull mode via WAP or MMS, and streaming.

OMA DRM v2.0 defines a suite of protocols called Rights Object Acquisition Protocol (ROAP). It contains three types of protocols:

- registration of user's devices;
- request and acquisition of rights object; and
- joining or leaving a domain.

To register a user device with a rights issuer, ROAP defines a four-pass protocol as illustrated in Fig. 7.11. This protocol is mandatory for the first connection of a device to a rights issuer. It includes mutual authentication, integrity protection, and certificate exchange. It may be also invoked to update some information, for instance the time of the device if the rights issuer deems it inaccurate.

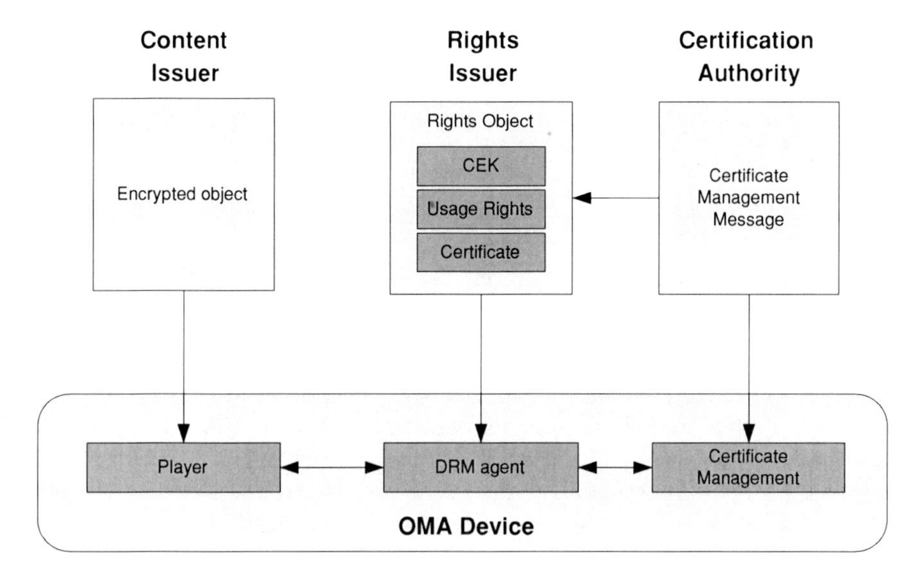

Fig. 7.10 OMA exchanges

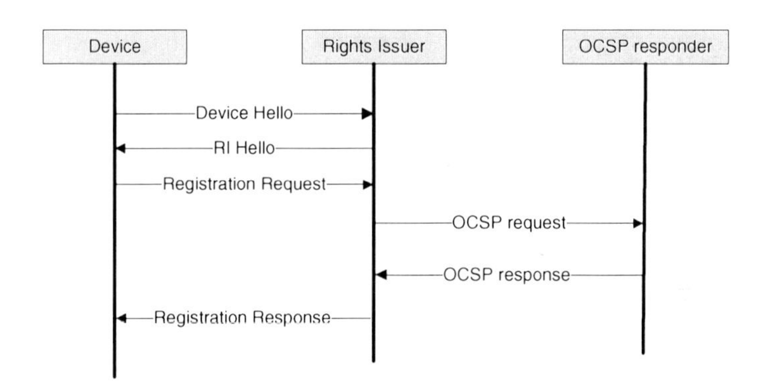

Fig. 7.11 ROAP device registration

The protocol may imply a third party: the Online Certificate Status Protocol (OCSP) responder. OCSP is an Internet protocol used to verify the validity of an X509 certificate [241]. It is an alternative to CRL. It uses an online server called the OCSP responder. The OCSP responder is optionally used to validate the rights issuer's certificate. The rights issuer sends the request with a nonce generated by the device. It forwards the OCSP response to the device. Thus, the device can verify that the rights issuer is not revoked. The device uses also the time in the response to synchronize its time.

The second protocol is used when a device acquires a rights object from a rights issuer. For that, the device must be already registered with the rights issuer. The protocol is a two-pass protocol. It includes mutual authentication and integrity protection. Once more, the rights issuer may optionally invoke an OCSP responder. A one-pass variant exists. It is mainly the second pass of the full protocol. It is done on the initiative of the rights issuer. It is typically used when the rights issuer regularly issues a rights object, for instance, in the case of subscription.

The third category of protocols is used to join or leave a domain. They are two-pass protocols. The device requests permission to join a domain from the rights issuer that handles this domain. In response, it receives the information to create a domain context. Without a domain context, the device cannot handle rights objects dedicated to a domain. The device requests permission to leave a domain from the rights issuer that handles this domain. Neither protocol ever invokes any OCSP responder.

OMA DRM is an exemplary illustration of the importance of the trust management layer for interoperability. Even if the DRM agent uses ODRL, supports all business models, and implements the standardized bulk encryption (AES), it is not sufficient for receiving content from any OMA merchant. The merchant and the DRM agent have to share the same trust layer. In other words, the rights issuer has to trust the manufacturer of the handset. For that purpose, they will use a common CA. The OMA consortium does not provide or manage any CA.[9] The OMA consortium only manages specifications and their updates, but OMA does not enforce compliance with these specifications. Content issuers, rights issuers, and manufacturers have to associate to set up a compliance and robustness regime (C&R). The C&R regime will provide its own CA. Currently, there is one available C&R regime: the Content Management License Administrator (CMLA) [242]. Intel, Panasonic, Nokia, and Samsung founded CMLA. CMLA enforces compliance with its adopted agreements through proprietary key distribution algorithms that devices must implement, acting as hook IPs (Sect. 3.8, Compliance and Robustness Rules).

Nevertheless, nothing prevents the establishment of other C&R regimes. The fact that OMA may have competing trust layers due to different C&R regimes may hinder OMA's interoperability. This contradicts the common belief that OMA is fully interoperable. If two C&R regimes are not compatible or do not want to interoperate,[10] then corresponding devices will not interoperate. Rights issuers may not trust devices complying with a given C&R regime. A device complying with C&R regime A may not be able to check the validity of a certificate issued by C&R regime B.

---

[9] This is a common issue with consortiums and standardization bodies. They often neither manage/define the C&R regime, nor manage the patent issues. Some companies have to set-up such a C&R authority or patent pool to license the involved patents. Without such infrastructure, the standard will never be implemented.

[10] Reasons why two C&R regimes may not want to be interoperable are many. For instance, the first C&R regime may require some constraints more restricting than a second one. In that case, rights issuers from the first C&R regime may not trust, at least for high-value content, the devices under the second C&R regime whereas the rights issuers affiliated with the second C&R regime may trust all devices from the first C&R regime.

Also, non-compatible C&R regime could be a strategy to protect a market, for instance, a regional market.

7.6 Marlin

Marlin is a DRM dedicated to consumer electronics devices initially designed by Intertrust, Panasonic, Philips, Samsung, and Sony in 2005 (Sect. 11.3.1, Coral) [243] under the umbrella of the Marlin Developer Community.

Marlin, like Coral, is based on two frameworks: Octopus DRM [244] and the Networked Environment for Media Orchestration (NEMO) [245]. Octopus is a generic engine that protects digital content regardless of the type of content (audio, video, text) and the usage rights. It uses two concepts: Octopus nodes[11] and links. An Octopus node represents a logical entity such as a user, a device, a domain, or a subscription. A link defines a logical relationship between two Octopus nodes. For instance, a link may declare that Octopus personality node A, which represents device A, belongs to Octopus user node Alice. The NEMO framework defines the secure communication between the different Marlin entities, the so-called NEMO nodes. A NEMO node is a trusted entity that securely interacts with others NEMO nodes. NEMO was initially designed by Intertrust, a US company that, like ContentGuard, patented many initial concepts of DRM. Since 2002, Philips and Sony have jointly owned Intertrust. NEMO's goals are multiple:

- Authentication: Every collaborating principal, called a node, has to mutually authenticate other nodes. This excludes, in theory, the possibility for pirate nodes to impersonate an authorized node. Only trusted nodes should interact together. For NEMO node B to authenticate NEMO node A:
  - NEMO node A has to prove to NEMO node B that it holds a secret data. This knowledge ensures that NEMO node A is an authentic device. Only authentic devices should hold this type of secret data.
  - NEMO node A has to prove to NEMO node B that it is identified by a given NEMO ID. This prevents impersonation between authentic devices.
- Message security, which ensures secure communication between collaborating nodes. NEMO enforces the privacy and the integrity of exchanged data.
- Authorization: Nodes may require peer nodes to have certain properties before communicating with them. NEMO declares these properties through attribute assertions. The authorization process is as follows:
  - NEMO node B provides all its attribute assertions to NEMO node A.
  - NEMO node A decides which attribute assertions it trusts.
  - NEMO node A decides if with these trusted assertions NEMO node B will be able to securely perform the expected action. If this is not the case, then NEMO node A cancels its request.

---

[11] Marlin has two types of nodes: nodes defined by the NEMO framework and nodes defined by the Octopus framework. To differentiate them, we will prefix them by the name of the framework they belong to.

- Service discovery: a node must be able to locate the nodes that may propose needed services. NEMO defines such a discovery service.

Marlin introduces two interesting concepts: binding and targeting.

- *Binding* means creating a cryptographic link between two items. Thus, a content key may be bound to a given device or to a set of devices (domain). In fact, the license carrying the content key is encrypted with the public key of the client node (or a secret key unique to the client node if a symmetric cryptosystem is used). The rights enforcement layer handles the binding.
- *Targeting* means defining the set of conditions the client node should present for the content object to be accessible. Targeting defines the set of mandatory Octopus nodes that must be reachable by links from the personality Octopus node. Two Octopus nodes are reachable if at least one path of links exists to connect both nodes.

Marlin uses a traditional DRM model. Figure 7.12 represents the Marlin system using the four-layer model, and Marlin's dedicated vocabulary. The piece of content to be protected is scrambled with a random key called the Content Key into a content object. The associated commercial rights are converted to a control program. The control program defines the targeting for a content object. It checks if the personality node has a set of links that reach the required nodes. The control program is executable code, written in Plankton byte code.[12] The content key is bound to the personality node. The control program is also bound to an Octopus node (not necessarily the personality node binding the content key). The Marlin license holds the encrypted content key, the encrypted control program, and the signature of the control program (symmetric using Hash-based Message Authentication Code (HMAC), or asymmetric).

The DRM client receives the Marlin license. If the license is bound to its personality node, the client decrypts the control program. It passes it to the Plankton virtual machine. The control program checks the targeting of the content object, i.e., whether the targeted nodes are reachable by this node. If this is the case, the DRM client decrypts the content key and then descrambles the content object.

Marlin uses different elements:

- The Marlin client is the software that will actually consume the content. Indeed, the client executes on the consumer electronics device.
- The registration service is the entity that provides the credentials for the Octopus nodes and links with the Marlin client.
- The domain manager is the entity that manages membership to a given domain. A domain manager may be remotely hosted on a server or may be a local service hosted by a Marlin Octopus node on the home network.

[12] Plankton is a dedicated basic virtual machine.

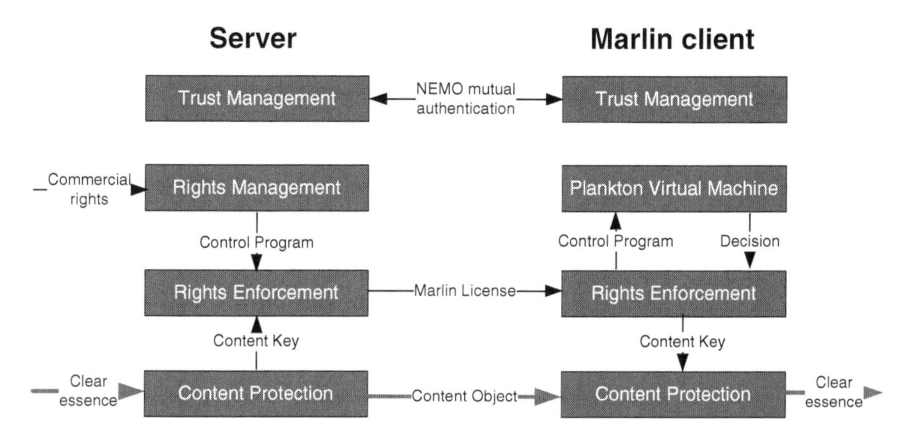

Fig. 7.12 Four-layer model of Marlin

- The license service is the entity that generates and delivers licenses to Marlin clients. The license's generation is triggered by a merchant request (Fig. 7.13).

When a consumer purchases a piece of content, the following actions occur once the merchant site has accepted the purchase:

- The Marlin client requests from the registration service the NEMO user node that identifies the customer. If the customer is known to the registration service, it will return a user node of the customer; else it will generate a new user node. Thus, the user node is associated with a given individual.
- Once the Marlin client gets the user node, it asks the merchant's registration service a link object. The registration service generates a link object that targets the NEMO user node to the client's Octopus personality node. In other words, it logically links the physical device to the consumer. The Octopus personality node is associated with a Marlin client, i.e., with a physical device.
- The Marlin client asks the license server for a license to use the purchased content object. For this purpose, Marlin client provides public information about the NEMO user node. The license server binds the license to the content object by encrypting the content key for the NEMO user node. Then the license server targets the NEMO user node by building the corresponding control program. It delivers the license to the Marlin client.
- The delivery of content the object can be done by any of the usual means.

Interestingly, the Marlin architecture separates license management from customer management. Access to the content by a customer is granted if the corresponding link object is bound to the NEMO user node. The content key is encrypted for that particular Nemo user node. The usage rights are defined for a given customer by targeting the NEMO user node.

Similarly, Marlin separates the management of content from the management of domain. Let us start with the easiest notion of domain: the user-based domain. The

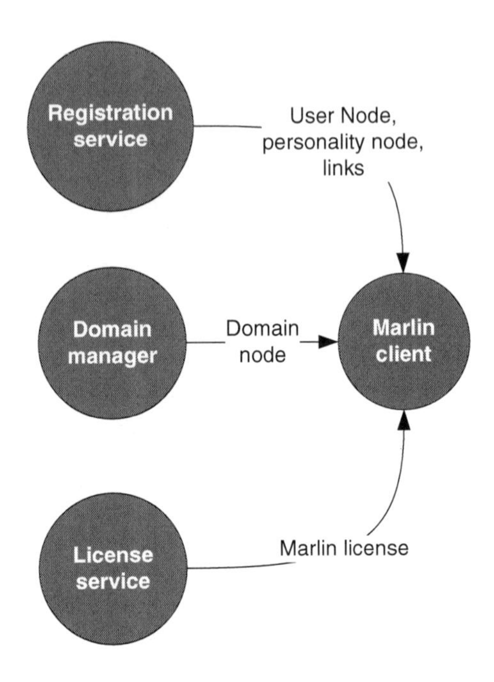

Fig. 7.13 Interaction between Marlin components

user-based domain encompasses all the devices that a user may own or use. The management of Alice's user domain is rather simple. It is done by linking the Octopus personality node of each of her devices to her user node. This is the task of the Domain Manager server. It manages some basic constraints such as the limit on the number of devices that Alice may own and the limit on the number of "owners" of a given device. In other words, Alice's user node may be linked to the personality node of her PC and the personality node of her STB. Thus, she can view her content objects on both her PC and her STB. Nevertheless, Bob's user node may also be linked to Alice's PC personality node. Thus, Bob can watch his content objects on Alice's PC.

Marlin introduces a more complex domain: the Marlin Common Domain [246]. In it, there is an additional node called the Octopus domain node. The personality nodes of the devices are linked to the domain node. The user nodes are linked to the domain node. Of course, the two types of domain may be combined.

In the example in Fig. 7.14, Alice and Bob share the same domain (of the family Doe), which encompasses the STB and the PC. Carol may share the Doe family's PC. Alice has her own phone. If a content object is linked to the Doe family's domain, Alice and Bob will be able to watch it on both the PC and the STB. Carol will be able to view her own pieces of content on Alice and Bob's PCs. If Alice purchases a content object for her phone, she will not be allowed to share it in the family's devices.

The Marlin Common Domain supports three mechanisms for the management of licenses and domains:

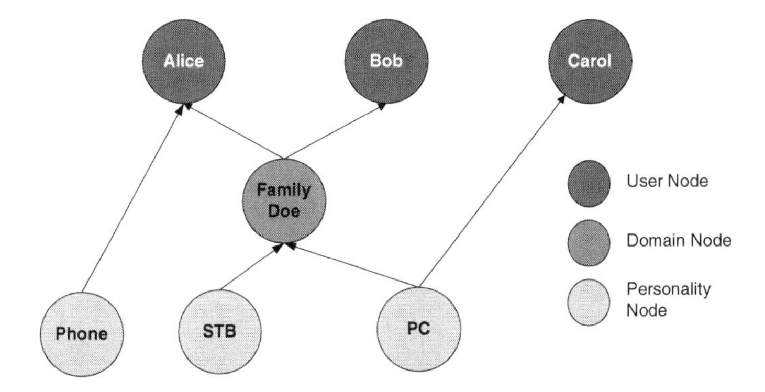

Fig. 7.14 Example of Marlin's domain configuration

- The basic management option targets and bounds the rights object to a user node. The content can be accessed through any device in the domains to which the consumer is linked.
- The releasable management option targets the user node but bounds it to the domain node. Only devices belonging to the bound domain can access the content.
- The releasable and moveable management option adds to previous mode the way to revoke a license or to transfer it to another user node. For instance, Alice could hand over her content object to Bob.

Being dedicated to consumer electronics devices, Marlin assumes that the DRM client executes in a controlled closed environment, and is thus more trusted than DRMs that execute on open computers. Thus, the trust assumption is more robust. For instance, the trust model may assume that hardware protects secret keys better than software obfuscation. Marlin uses a set of specifications:

- The Marlin Core System Specification defines the different common mandatory elements and their interactions at a high level.
- The Marlin Broadband Delivery Specification defines additional capacities of Internet-enabled devices, such as online license retrieval.
- The OMArlin Specification defines the interoperability between OMA and Marlin, in other words how to share content with OMA-compliant mobile phones.
- The Marlin IPTV End-point Services Specification defines the behavior of IP TV sets.
- The Marlin Common Domain Specification defines the implementation of the authorized domain, i.e., the management of home network devices.
- The NEMO (Networked Environment for Media Orchestration) framework.

A single central authority manages the trust: the Marlin Trust Management Organization. It publishes Marlin's C&R Regime. This organization can delegate the generation of Marlin certified keys to a third party. Currently, Intertrust manages this role under the name Seacert [247].[13]

Marlin is currently deployed on Sony's PlayStation 3 (PS3) and PlayStation Portable (PSP), is part of the Japanese national IPTV deployment, and will be one of the mandatory DRM systems for BBC's YouView.

7.7 Some Other Players

We have already presented some DRM solutions that are market leaders or are expected to become dominant, and others for educational purpose. Many other solutions are competing in this field. All solutions use the same architecture, sometimes with their own specific vocabulary. We will not provide an exhaustive list of the existing solutions. Nevertheless, we will list some of the most prominent ones.

Widevine provides a DRM for numerous platforms ranging from Windows OS, and MAC OS to Linux-based consumer electronics devices (called Cypher Widewine). The DRM client uses the Cypher Virtual smart card, which emulates a virtual smart card through purely software implementation. Widevine provides a session watermark with its Mensor solution. It embeds a 64-bit payload either at the head end or at the terminal. Widevine has licensed watermark technologies from three undisclosed suppliers [248]. Furthermore, Widevine claims that it monitors the potential activity of video scrappers, i.e., pieces of software that capture video displayed on a computer screen. Once a video scrapper is detected, the system may terminate the playback, degrade its quality, or silently monitor and report the activity. On December 2010, Google acquired Widevine [249]. This event may boost Cypher Widevine to a future key player. We can reasonably expect to see it protecting Google TV and even some copyrighted content on YouTube [250].

Verimatrix offers Video Content Authority System (VCAS3), which supports IPTV and OTT delivery. Asymmetric cryptography protects licenses using X509 certificates. VCAS supports Windows OS, MAC OS, and Linux-based consumer electronics devices. VCAS can also support DVB-compliant STB with detachable hardware components, i.e., smart cards. Verimatrix has its own watermark technologies: VideoMark has been specifically designed for embedding into terminals whereas StreamMark has been specifically designed for embedding content of the head end into the compressed domain.

SecureMedia offers Encryptonite, a CAS and DRM solution that supports Windows and Linux solutions. The preferred scrambling algorithm is RC4, but it can be replaced by AES or 3DES. Furthermore, Encryptonite can use DVB-CSA with Simulcrypt. In January 2010, Motorola acquired SecureMedia, strengthening its offerings in content protection after the acquisition of General Electric's division.

There are many other solutions, such as Real's Helix and Medialive (recently acquired by Nagravision).

---

[13] Seacert generates also keys for OMA and OMArlin.

Chapter 8
Protection of Pre-recorded/Recordable Medium

The protection of pre-recorded and recordable media is highly dependent on the characteristics of the media itself. We will take a brief tour of the protection methods of the most important types of media, mostly optical media. The tour will be organized in the order of their introduction into the market. Thus, Sect. 8.1 provides an overview of the methods that prevent the duplication of Compact Discs (CD) and other optical media. Section 8.2 describes how DVDs protect video. Section 8.3 details the protection method of Sony's memory stick. Section 8.4 describes a framework used to protect recordable DVDs and some memory cards. Section 8.5 introduces the protection scheme used for Blu-ray discs.

8.1 Anti-Ripping

Designed in 1980 by Philips and Sony, the Red Book, also known as IEC809 is the standard specifying CD audio (CD-A) [251]. It defines the physical characteristics (dimensions, characteristics of light reflection, size of the pits, or coding principle), the logical organization on the disc, and the way sound is digitized (Pulse Coding Mode, 44.1 kHz sampling frequency). Unfortunately, the specifications did not define any copy protection system. Audio tracks are stored in the clear on the disc. At the time of design of CD audio, digital piracy was not a concern for content owners. In the 1980s, piracy was restricted to analog copies of audio tapes. Successive copies of an audio tape degraded the audio quality. Nobody anticipated that, a few years later, burning CDs would become a common, cheap, and easy task. Recordable CDs were still in the laboratories. When introduced first in the market, recordable discs were rather expensive items. Unfortunately, with the advent of personal computers, and cheaper recordable CD-R discs, it became easy to rip any audio CD. Furthermore, successive copies of a digital track still produce a pristine copy. In fact, copying a digital track creates a clone. Thus, several companies attempted to add copy protection to audio CDs to thwart the duplication

E. Diehl, *Securing Digital Video*, DOI: 10.1007/978-3-642-17345-5_8,

of CD-A. The same techniques were extended to protect other media such as CD-ROM (Read Only Memory) or DVD. They used several families of techniques:

- using CD-ROM dedicated tracks to fool the ripping computer;
- spoiling the Table Of Contents (TOC) of the file system of this disc;
- adding errors to some data; and
- taking control of the computer during play back.

The first type of technique uses the difference in behavior between a CD-A player and a computer that is able to play CD-A, and software stored on a CD-ROM. The CD-ROM standard introduced a new concept: multi-session discs. Conceptually, a multi-session disc is a set of stored virtual discs on the same physical disc. A CD-A normally has only one session. When a computer drive reads a multi-session disc, it first looks for the last session, whereas the regular CD-A player only reads the first session.[1] If a computer drive finds tracks containing software, it automatically executes them.

Thus, the main idea of this type of protection is to create a multi-session disc whose first section contains the audio tracks whereas another session contains a crafted piece of software. A CD-A player will directly jump to the first session, i.e., the audio tracks, whereas a computer will first attempt to execute the software of the second session. The goal is for this piece of software to thwart any attempt to duplicate the disc.

Two instances of this technique made headlines in the press:

- Sony's Key2Audio: This solution, designed in 2001 by Sony DADC [252], added a bogus data track on the CD-Audio. When reading the disc, the computer accessed the bogus file, was trapped in an endless loop, and could not escape to play the audio files. Unfortunately, the bogus data track was physically located on the outer ring. Blacking out the outer ring of the disc with a pen was sufficient to defeat the system [253]. The computer did not anymore detect the presence of the bogus track, and directly jumped to the audio tracks.
- First4Internet's XCP: In November 2005, Sony BMG sold two million music CDs (about 50 titles) embedded with a copy protection system, called Extended Copy Protection (XCP). Unfortunately, XCP used a persistent rootkit. (For more details, see the next "The Devil's in the Details"). A security expert discovered the rootkit. A few days after the rootkit disclosure, some crackers exploited the stealth ability of the already installed rootkit to install their own hidden "malware" (similar to a Trojan horse). Facing this disaster, Sony BMG provided a patch to uninstall the rootkit [254]. Unfortunately, the patch came with its very own problems. The first version of the patch was written too fast. It generated stability problems on computers to which it was applied. Furthermore, this patch created a new security flaw whenever the customer visited websites controlled by crackers. The crackers installed malicious programs using the

[1] In fact, a CD-A player is not even aware of the existence of other potential sessions and thus does not look for them.

resident Sony BMG deactivation tool. Later, Sony delivered a proper patch to address the problem.

Sony BMG, to counter piracy, employed the very same malicious techniques as pirates did, such as cloaking, rootkits, and spyware. In the same spirit, the End User License Agreement (EULA) did not mention that XCP installs hidden files on a user's computer. Furthermore, the installation, obviously, did not provide tools to uninstall the rootkit. XCP was not ethical.

Soon, in addition to the technical and reputational problems, came legal ones. Several lawsuits were launched against Sony BMG. The Electronic Frontier Foundation (EFF) attacked Sony for endangering consumers through a flawed copy protection system [255]. The negotiated settlement allowed each owner of XCP-protected CDs to choose for every returned CD to receive either a cash payment of $7.50 and one free download from a list of approximately 200 titles in the Sony BMG catalogs, or three free downloads from this list.

The Devil's in the Details

When executed for the first time on a Windows system, XCP opens a dialog box to accept the EULA. Then, it installs three pieces of software: a new driver for the CD/DVD drive that replaces the current one, a Sony-dedicated audio player, and a stealthy driver. Unfortunately, XCP software is intrusive. To remain hidden from the user, and intercept all copy requests or playback actions, XCP uses a rootkit designed by the company First4Internet.

A rootkit is a piece of software that hides itself from the user. It operates very closely to the kernel of the operating system (usually at least in Ring 0, where it has access to the entire memory space and to the complete instruction set). It permanently monitors communications between users, and applications and this kernel and between the different components of the kernel. Thus, the rootkit can intercept queries that may disclose its presence, and replace potentially compromising data with an innocent fake answer. It acts like a man in the middle attack [256]. For instance, if the rootkit sees a request to list all files in the directory it is currently stored in, it will carefully erase its presence in the returned answer. The XCP rootkit intercepted all calls requesting access to the CD drive, and thus was able to control the number of copies.

Crackers use a rootkit to hide a process or an open point that controls the infected computer. For instance, XCP's rootkit hides all files whose name contains the pattern "`$sys$`".

A rootkit can only intercept a set of predefined queries. Thus, there are always some specific queries that will detect a rootkit. Its response does not match the expected one. Rootkit detectors exploit this difference of behavior. Rootkits and their detectors are now arm wrestling, like virus and antivirus writers. Rootkits have derived from piracy techniques used by malware writers. They alter the core operating system, degrade performance, and increase instability. According to malware experts, XCP may also act as a

spyware application by connecting to online sites during a playback sequence (which may lead to user privacy issues).
On 31 October 2005, Mark Russinovich, a rootkit expert, discovered a new rootkit installed on his PC [257]. After investigation, he identified the source: the XCP copy protection of Sony BMG. A detailed analysis also highlighted that XCP infringes the Gnu Public License (GPL). It includes a portion of source code written by the (in)famous hacker DVD Jon under the GPL. First4Internet violated the legal GPL agreement that would have obliged them to publish their source code.
Lesson: Any system, legal or pirate, has weaknesses and somebody will discover them.

The second family of techniques spoils the TOC [258]. A disc is composed of a set of tracks. The tracks may contain audio, video, subtitles, or data. The TOC is a specific track that lists all the tracks available on the disc. It details their type, size, and location. This anti-ripping technique modifies the TOC read by the computer using tricks such as changing the start time (i.e., the position in the disc), the length, or the type of stored information. Thus, some protected discs may declare data larger than the actual capacity of the disc, or announce ghost tracks, i.e., nonexistent tracks. In theory, this should fool the copying software. Unfortunately, sometimes it introduces incompatibility with regular CD players. Using the trick of the multi-session, the authoring of this disc spoils the sessions that the computer will access first. This technique is sometimes associated with the automatic execution of control software, as is the previous technique.

The third technique exploits the error correction mechanism of the drive. Physical damage to the surface of the disc, such as scratches, may corrupt the read data. To protect against such damage, the media uses additional data in the form of error correcting symbols. An Error Correcting Code (ECC) not only detects the presence of bogus data, but, under some circumstances, can even correct the errors. Often, the longer the correcting code, the more efficient the scheme. Each sector of the disc has its own ECC data.

Some copy protection technologies (like Rovi's[2] RipGuard [50] or Sony DADC's ARccOS [259]) intentionally fill, at the authoring stage, sectors with corrupted data and incorrect ECC blocks. Usually, when a drive encounters a bogus section that it cannot correct with the retrieved ECC data, it attempts to read it again until it recovers valid data. In order to prevent a lock condition, after a given number of trials,[3] the drive flags this sector as a "bad" sector and skips it. Video and audio players skip any "bad" sector and thus are not impaired by this bogus data. Some DVD ripping programs are thwarted since they attempt to copy all sectors of the disc, including the bad sectors. In this case, they stay blocked in

[2] Since 2009, Rovi has been the new name of Macrovision.

[3] The number of trials depends on the model of the drive.

an endless loop. Indeed, DVD ripping programs try as much as possible to avoid altering the structure of copied disc because some copy protection systems may check the presence of bad sectors. Thus, the presence of bad sectors may be relevant.

For instance, some copy protection methods alter the sector numbers.

Normally, the sector numbers on a disc are sequential. These copy protection systems may invert the order of some sector numbers, duplicate some sector numbers, or drop some sector numbers. Such alterations do not affect the normal playback of an original disc, as the player only needs to address the sectors starting a chapter or an audio track. Once it has found the starting data, it reads the other sectors in their physical order on the disc, regardless of their numbers. However, a DVD recorder generally will perform a copying operation by accessing the sectors in a logical, sequential order. It does this systematic access to copy all sectors, even sectors not referenced by the table of contents. Missing sector numbers, sector numbers in a non-sequential order, and duplicate sector numbers all give difficulties to DVD ripping software, and may cause it to fail while reading or copying [260].

The fourth family of techniques installs control tasks during the playback of a disc. Thus, it is similar to the trick used by Sony with XCP. When the disc is launched, the software executes and attempts to take control of the computer. Once the computer is under the control of this software, it should be possible to discard any attempt to copy a disc. For instance, some systems install a dedicated media player to replace the resident one, at least to play back the inserted disc. Of course, this dedicated media player forbids any copy.

In 2003, John Halderman described such a solution developed by SunnComm's MediaMax CD3 system [261]. SunnComm used the `autorun` feature of Windows. The `autorun` feature, if switched on,[4] automatically launches the executable files listed in the file `autorun.inf` stored in the root directory of a removable medium. When loading the MediaMax CD3 protected CD, Windows executes the application `launchCD.exe`. This application displays the EULA. If the user accepts the EULA, the application installs a driver `SbcpHid` that remains active even after a reboot. This driver is not stealthy. It intercepts all calls to the drive and attempts to identify the inserted disc. If the disc is supposed to be protected by MediaMax CD3, the installed driver blocks any attempt to duplicate this disc. If the user does not accept the EULA, reading of the disc is terminated.

Unfortunately for SunnComm, their copy protection system can be easily bypassed without requiring any technical skills. There are two solutions:

- Deactivate the `autorun` feature from Windows; this should already have been done on all computers of security-aware users. The `autorun` feature is one of the most used vectors of contamination by viruses [262].
- Keep the keyboard's Shift key pressed down while launching the disc. This action temporarily disables the Windows `autorun` feature, thus preventing the

[4] By default, the `autorun` feature is active in Windows distribution.

Fig. 8.1 Normal CD-A and copy controlled CD-A logos

application `launchCD` from executing. The simplicity of this "circumvention" method made headlines in the media [263].

The advantage of this last family of techniques is that it does not introduce, if used alone, any non-compliance issue with CD players. The main disadvantage, and it is a serious one, is that it often adds drivers to the host system. The most benign problem with this is that installed drivers clog the resources and compromise performance of the computer, as does spyware. The worst problem is that sometimes they introduce instability in the host system.

In fact, most current systems are a combination of the four described families of techniques. Although we mostly described the techniques with CD-A as target, most of them contribute to preventing ripping of CD-ROMs and DVDs. Of course, there is an endless arms race between these protection schemes and the newest versions of the pieces of ripping software. The latter always learn the new deployed tricks and implement corresponding "countermeasures".

Can these systems efficiently prevent ripping? Any CD-A has to comply with the Red Book standard. This means that any copy protection scheme for audio CDs is doomed for defeat. In a publicly known, precise location on the disc, the audio tracks are in the clear. The hacker has just to look for this clear data, jumping over any protection. Mandatory support of legacy players killed the chance of any serious protection. Of course, no content owner is ready to lose some of his potential markct because some consumer devices cannot play his content. Content owners cannot afford not to be compliant with 100 % of the installed base. Every launch of title using non-fully compliant solutions turned out into a disaster. Each time, the content providers had to release the same title without protection [264], after violent fire storming of media. Furthermore, consumer associations complained that CD-A protected against copying should not display the CD logo because it does not respect the standard. To circumvent this restriction, some distributors, such as EMI or Sony, introduced a new logo to label protected CD-A (Fig. 8.1).

If all such anti-ripping systems are doomed to fail, why does the media industry still use them? The media industry is clearly aware of this limitation. Its objective is to prevent ordinary people from breaching security, i.e., keeping honest people honest. These systems target 95 % of the users. According to Rovi, one of the major providers of such technology [50]:

> While no encryption method is 100 percent foolproof, the goal of a good security solution is to deter the majority of violators. Between 85 to 95 percent of all consumers lack the patience and the technical know-how to break through the extra layer of DVD copy

protection Rovi's RipGuard provides, causing them to give up on making illegal copies. Often, casual users end up purchasing the DVD content they were trying to copy.

Therefore, these techniques are deployed for many titles, and they have some impacts on the sales, although it is difficult, if not impossible, to evaluate the actual positive impact.

8.2 CSS: Content Scramble System

Video piracy existed before the advent of DVD. Illegally copying videotapes on a small scale was not economically viable to become widespread. Studios fought copying on a large scale through legal measures. It was rather difficult to hide farms of VHS recorders. Studios did not worry about casual copies of videotapes for friends (so-called ant piracy).[5] Video piracy was rather under control. When the movie industry decided to replace the VHS tape with the DVD, it wanted to avoid the past mistakes of the audio industry. The misfortune of the Recording Industry Association of America (RIAA) was a canary in a coalmine. Content protection had to be part of any future video format standard, and that from the start. Thus in 1996, MPAA launched the CPTWG [101]. This working group is a coalition of major studios, consumer electronics manufacturers, and information technology providers with the goal of preventing illegal duplication and redistribution on Internet of digital content. Since then, every 2 months, technical experts and lawyers from major studios, consumer electronics manufacturers, and information technology (IT) industry representatives have met to design solutions for protecting digital content. Fifteen years later, the CPTWG is still active [265]. Nevertheless, the CPTWG does not anymore design new security schemes.

Its first outcome was the Content Scramble System (CSS) designed by Matsushita. This standard defines the encryption method of DVDs. CSS uses three 40-bit keys:

- The title keys protect the actual content. The content is scrambled through a proprietary symmetric cryptographic algorithm using the title key. This algorithm has never been published. The same disc may use different title keys for different sections, or use one single title key for all sections.
- The disc key protects the title keys. The disc key is unique for a disc, i.e., for one given movie title there is one single disc key. The title keys are encrypted using the disc key. The disc holds the encrypted title keys.
- Player keys are unique for each manufacturer. Every player of a manufacturer holds the player key of this manufacturer. The manufacturer receives its key

[5] It should be noted that already with VHS tapes, there was an attempt to use some analog content protection. In 1983, Macrovision designed an analog copy protection system called APS. Its objective was to impair the behavior of the automatic gain control of the analog input to the VHS recorder.

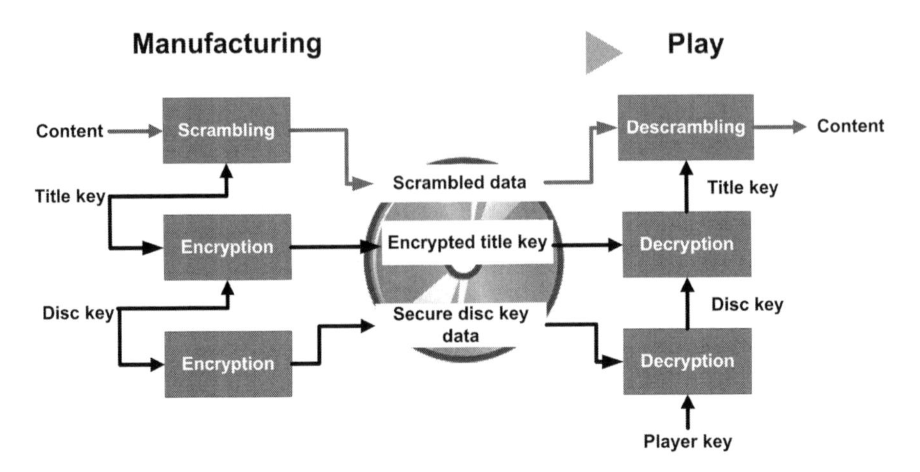

Fig. 8.2 CSS protection

from the DVD Copy Control Association (DVD-CCA) once it signed the license agreement. CSS can support up to 409 manufacturers. A data structure, called secured disc key data, contains the disc key encrypted by all 409-player keys. In other words, the secured disc key data contains 409 encrypted disc keys.

Figure 8.2 illustrates the CSS process for protecting a DVD and playing it back. When building the glass master[6] for the DVD, the manufacturer scrambles the different sectors with the corresponding random title keys. A random disc key encrypts the title keys. These are the encrypted title keys. DVD-CCA encrypts the disc key with the 409 manufacturer keys creating the secured disc key data that is stored in the Lead-In Area.[7] All this data is pressed into the final DVD.

The DVD player retrieves the secured disc key data from the Lead-In Area of the DVD. Using its player key, it extracts the disc key by decrypting the proper encrypted key. In theory, it is possible to revoke all the players from one given manufacturer. In such a case, the secured disc key data would not anymore contain the disc key encrypted by the corresponding manufacturer's player key.

Using the disc key, the player decrypts the encrypted title keys from the DVD. With the title key, the DVD player can descramble the protected content of the DVD.

[6] The glass master is the master stamp used to press a DVD or CD. It is a negative image of the actual disc.

[7] The Lead-In Area is in the inner part of the optical disc. Standard consumer DVD recorders cannot write on the Lead-In Area. A special laser is required to write on this area. Of course, when pressing a disc from a glass master, this is not a limitation. Thus, it is impossible to duplicate the CSS-encrypted disc through a bit-to-bit copy. There is no way to write the secured disc key on the duplicated disc. Therefore, copies of a DVD are not encrypted (at least with a generic DVD recorder).

Table 8.1 CSS region coding

Region code	
0	Playable in all regions
1	Bermuda, Canada, United States, and US territories
2	The Middle East, Western Europe, Central Europe, Egypt, French overseas territories, Greenland, Japan, Lesotho, South Africa, and Swaziland
3	Southeast Asia, Hong Kong, Macau, South Korea, and Taiwan
4	Australia, New Zealand, Central America, the Caribbean, Mexico, Oceania, and South America
5	The rest of Africa, former Soviet Union, the Indian subcontinent, Mongolia, and North Korea
6	Mainland China
7	Reserved for future use
8	International venues such as planes, cruise ships, etc.

It is interesting to highlight the strong coupling between the technology used (encryption) and legal enforcement through licensing. If a manufacturer wants to build a DVD player that is able to play the content from major studios, it will need to get a manufacturer key. For that, the manufacturer has to sign the license agreement with the DVD-CCA. Of course, a signature on the license agreement implies acceptance of the compliance and robustness regime (Sect. 3.8). This allows, for instance, imposing regional codes. Region codes identify the zone of the world where a given DVD can be played. A DVD player should be attached to one single region code. The DVD-CCA sliced the world into nine regions (Table 8.1).

In October 1999, 3 years after the first DVD was introduced in Japan, Jon Lech Johansen, nicknamed DVD Jon, member of a team of Norwegian hackers called Masters of Reverse Engineering (MoRE), published DeCSS, a piece of software that circumvented CSS protection. How did he succeed? A subsidiary of Real Networks, XING Technology, did not properly protect its player key in a software-based DVD player. With XING's valid player key, it was possible to rewrite an application that implemented the secret specifications [266]. In fact, DVD Jon reverse-engineered the complete software to rebuild the decryption process. Furthermore, using XING's player key, he and his team reverse-engineered the player keys of 170 manufacturers. In fact, with the XING's player key, DVD Jon had access to the pair {encrypted disc key, clear disc key} for each manufacturer key. The manufacturer key was 40 bits long. At this time, 40-bit keys were already far too easy to break by brute force. DVD Jon could brute-force all manufacturer keys. His software, DeCSS, implemented all keys.[8] As a nice analogy, DeCSS was not a

[8] Once the first version of DeCSS was published by DVD John on the 2600.org forum, the community was extremely creative in disseminating it. First, it was modified, enhanced, and ported to many languages such as PERL and Java. The best example is most probably the PERL

lock pick but rather a ring of copied keys from somebody who left his keys on the door. It was impossible to revoke DeCSS without revoking many legitimate DVD players in the market.

One of the design errors was to use a too small key size. The cryptography community knew that 40-bit keys were too short to be secure. Nevertheless, using 40-bit keys drastically simplified the problem of exporting (or importing) cryptography. In many countries, cryptography is considered a war weapon. As such, its use and sale require specific authorization from government agencies [268]. The design decision was mainly driven by US regulation of cryptographic algorithms. All countries accepted importation of 40-bit key-based cryptography. At the end of 1990s, many countries were still banning export or import of 64-bit and even 56-bit cryptosystems. Using a key size larger than 40 bits would have required seeking authorization for CSS in many countries around the world. This solution would have been too risky and far too slow for fast market adoption.

Once the encryption algorithm was disclosed, serious cryptanalysis proved that the algorithm itself was weaker than even 40 bits [269]. Furthermore, the key management method was too simple. It only used one stage of player keys, thus offering no good granularity. The system was not renewable. The response to a hack was revocation. Unfortunately, revocation used a limited set of global static keys, which were all compromised.

The second error was to allow software implementation of the DVD software player [270]. As mentioned, tamper-resistance of software is limited, and if not applied at all, is even easier to crack. This is unfortunately often the case. Furthermore, when the goals of the different actors do not align, security fails. In general, if the designer of the secure system is not the one who will suffer if it is broken, problems can often be foreseen [271]. The designers of the DVD player have no strong incentive, except goodwill or the need to comply, to make a robust implementation. They do not suffer actual loss when it is compromised. The content owners incur the loss. A robust implementation is even at odds with their commercial incentive. It requires more complex development and hence is more expensive.

Studios learned from the first error and enhanced subsequent schemes. Unfortunately, they did not take into account the second error, as we will see later with AACS[9] (Sect. 8.5.1).

(Footnote 8 continued)
implementations which were rather tiny. Second, surprising supports were used to disseminate the hack. CSS descrambler source code was printed on T-shirts, ties, used to compose a dramatic reading on radio, and presented as a Haiku poem! The Gallery of CSS descramblers is a homage to creativity [267]. The DeCSS story also led to vigorous litigation that invoked the US First Amendment, free speech, and protected software code.

[9] Either this or they were not able to implement a solution acceptable of all stakeholders that would correct the second error.

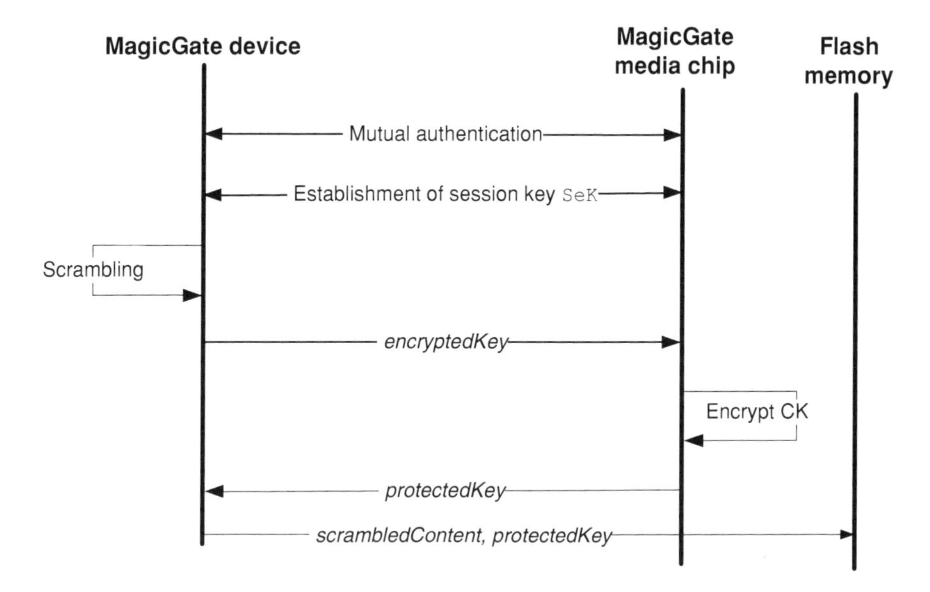

Fig. 8.3 Recording MagicGate-protected content

8.3 MagicGate

In 1999, Sony proposed MagicGate [272]. MagicGate is dedicated to memory cards, such as Sony's Memory Stick, and multimedia cards. MagicGate employs a microchip embedded in both the player/recorder and the actual media to ensure two functions [273]: mutual authentication and protection of content (Fig. 8.3).

The processor of the player and the processor of the media mutually authenticate each other. Through this method, only compliant devices can access MagicGate-protected media. As usual, the successful authentication process generates a session key, `SeK`.

Authentication through a physical component prevents unauthorized duplication of stored content. It is always possible to replicate the stored content, but without the chipset (whose distribution is controlled),[10] it is useless.

When recording a piece of content on a MagicGate media, for instance, storing it in a memory stick, the following process occurs:

- The MagicGate device chip scrambles the content with a proprietary undisclosed algorithm, hereafter called E1, using a random content key `CK`. We get $scrambledContent = E1_{\{CK\}}(content)$.

[10] The security assumption is that crackers will not be able to mimic the behavior of the microchip.

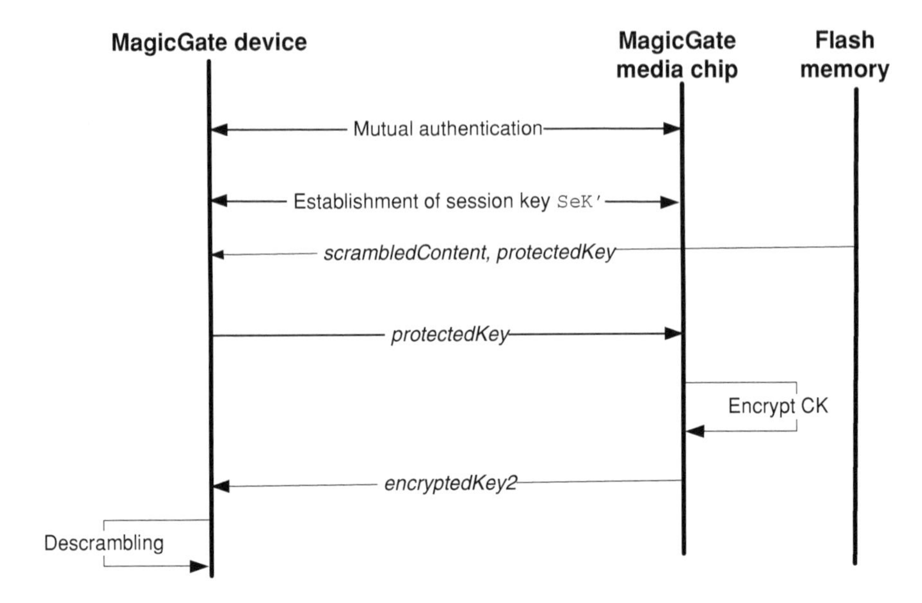

Fig. 8.4 Reading MagicGate-protected content

- The MagicGate device chip encrypts the content key `CK` with a proprietary undisclosed algorithm, hereafter called E2, using session key `SeK` such that $encryptedKey = E2_{\{SeK\}}(CK)$. It sends the *encryptedKey* to the MagicGate media chip.
- The MagicGate media chip decrypts *encryptedKey* with session key `SeK` to retrieve `CK`. The session key is established during the authentication phase. Therefore, both chips know `SeK`. Then, it encrypts content key `CK` with a proprietary undisclosed algorithm, hereafter called E3, using its unique secret key, the storage key `Kstm`. Each MagicGate media chip has such a unique key. The result is $protectedKey = E3_{\{Kstm\}}(CK)$.
- The MagicGate media chip returns protectedKey to the MagicGate device chip.
- The device stores *scrambledContent* together with *protectedKey* in the flash memory of the memory stick.

When playing back a piece of protected content from MagicGate media, for instance, reading it from a memory stick, the following process occurs (Fig. 8.4):

- The device reads both *scrambledContent* and its *protectedKey* from the flash memory of the memory stick.
- The MagicGate device chip sends *protectedKey* to the MagicGate media chip.
- The MagicGate media chip decrypts *protectedKey* with its storage key `Kstm` to retrieve `CK`. Then, it encrypts content key `CK` with the new session key `SeK'`. The result is $encryptedKey2 = E2_{\{SeK'\}}(CK)$.
- The MagicGate media chip sends *encryptedKey*2 to the MagicGate media chip.

- The MagicGate device chip decrypts *encryptedKey*2 using session key `SeK'`[11] such that $D2_{\{SeK'\}}(encryptedKey2) = CK$.
- The MagicGate device chip descrambles the content using retrieved content key `CK`.

Thus, the content is protected during transmission and storage. Only compliant MagicGate devices, i.e., those equipped with a MagicGate chip, can descramble MagicGate protected content, thus preventing illegal playback. Furthermore, the encryption of content key `CK` with a key unique to the storage media prevents unauthorized transfer of files from Alice's memory stick to Bob's memory stick. Bob's MagicGate media chip will not be able to retrieve the right `CK` from *protected* key because it has a different storage key, `Kstm`.

8.4 CPPM CPRM

One of the numerous critiques about early content protection systems was that they only solved isolated problems such as protecting a given type of medium, and protecting a given link. It was a piecemeal solution. Manufacturers and content owners rallied for a more global view of content protection. Thus, IBM, Intel, Matsushita, and Toshiba, the so-called 4C Entity, created a larger framework of content protection: Content Protection System Architecture (CPSA). It attempted to combine several technologies, such as CSS and DTCP (Sect. 9.3), in a unified approach. For that, they had also to close some holes. Several elements were not protected.

In October 1999, 4C Entity introduced the Copy Protection for Recordable Media (CPRM) system. In April 2000, following the exploit of DVD Jon, who broke CSS, 4C introduced the Copy Protection for Prerecorded Media (CPPM) system. Both schemes are more sophisticated than CSS and introduce interesting new features:

- Media type recognition: The media drive should be able to distinguish a non-recordable medium from a recordable one. The presence of CSS-protected or CPPM-protected content on a recordable medium should be treated as an illegal use, and thus the media drive should abort playback.[12]
- Media ID: Each recordable disc has a media ID. In fact, the media ID is not unique per disc. Several discs may have the same media ID. The encryption of the data on the disc uses this ID. The first assumption is that the media ID of a recordable disc cannot be forged or modified. The second assumption is that the probability of finding two recordable discs with the same media ID is extremely low, which makes the media ID quasi-unique. With these assumptions, bit-to-bit copy is impossible. Bit-to-bit copy is the attempt to make a perfect, identical

---

[11] In this explanation, we assumed that the playback occurred in a different session, i.e., the memory stick was unplugged and plugged in again, or the device was turned off.

[12] The CSS and CPPM license agreements clearly exclude the use of these protection schemes on recordable media. CSS and CPPM should exclusively be used for pressed media.

copy of a disc, with the same organization, the same sector numbers, and the same ECC values.[13]

- Media Key Block (MKB): this data structure manages the revocation of rogue or compromised players. We will describe more in detail the MKB because it has shaped the new generation of pre-recorded content.

MKB uses broadcast encryption. Broadcast encryption is a new type of cryptosystem that was first presented by Amos Fiat and Moni Naor in 1993 [274]. The aim was to encrypt messages that could only be received by a subset of recipients within a known population. The additional strict constraint was to only use one-way transmission; hence the term broadcast encryption. In the initial papers, the first targeted application was Pay TV, although no commercial Conditional Access System currently employs it. In 1997, Brendan Traw from Intel had the idea to apply broadcast encryption to optical media [275]. We will present a simplified explanation of broadcast encryption.

CPRM (as well as CPPM) uses a basic form of broadcast encryption. We will use a simplified example to illustrate this initial implementation. The authority generates many random keys $K_{i,j}$ that it stores in a matrix. This simplified example arbitrarily defines the number of columns as four. The authority distributes to each device a set of four of these keys, called device-keys, such that each device has one of its device-keys in each column. Two devices may share some device-keys, but no two devices will share all four device-keys. We will assume that Alice's device-keys are $K_{1,1}$, $K_{2,2}$, $K_{2,3}$, and $K_{3,4}$ whereas Bob's device-keys are $K_{1,1}$, $K_{1,2}$, $K_{3,3}$, and $K_{2,4}$. The reader may see that Alice's device and Bob's device share one device key, $K_{1,1}$.

The authority builds the MKB by defining a random media key K_{media}. The MKB is the matrix containing this media key, K_{media}, encrypted with all the keys of the initial matrix. Thus, the element of the ith row and jth column of the MKB is $C_{i,j} = E_{\{K_{i,j}\}}(K_{media})$. MKB looks like the matrix on the left-hand side of Fig. 8.5. In this figure, the values encrypted with Alice's device keys are circled in blue, whereas those encrypted with Bob's device-keys are circled in green. Alice's device uses its first device-key $K_{1,1}$ to extract the media key $K_{media} = D_{\{K_{1,1}\}}(C_{1,1})$. Similarly, Bob's device can extract the media key using the same device-key. In fact, Alice and Bob's devices could use any of their four device-keys to retrieve the media key.

Let us assume now that the Authority needs to blacklist Bob's device. It then issues a new MKB using a new media key. Every location of the MKB that would have been encrypted with Bob's device-keys will be filled with a random value instead of the expected value. In other words, the generation of the new MKB follows the following algorithm:

[13] Some copying software just replicates the logical organization of the original disc. For instance, DVD rippers that copy CSS-protected discs first decrypt the CSS-protected data before burning it to the final disc. Thus, they may modify the data.

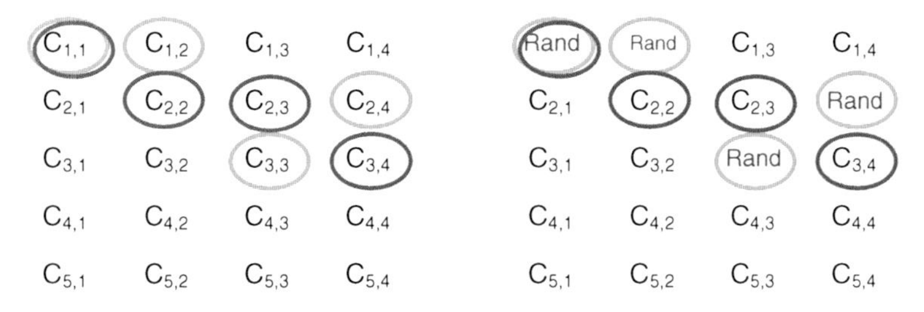

Fig. 8.5 Broadcast encryption for CPRM

If $K_{i,j}$ is not revoked

  then $C_{i,j} = E_{\{K_{i,j}\}}(K_{media})$

  else $C_{i,j} =$ random()

end if

The right-hand side of Fig. 8.5 illustrates the new MKB with the random values highlighted in red. When Alice's device attempts to extract the media key, it first uses its first device key $K_{1,1}$ which unfortunately is also one of Bob's device-keys. Hence, Alice's device cannot decrypt $C_{1,1}$. It then uses its second device-key $K_{2,2}$ to decrypt $C_{2,2}$. Alice's device, not being revoked, it succeeds in retrieving the new media key. On the other hand, Bob's device keys being all revoked, every attempt to decrypt the corresponding values (i.e., $C_{1,1}$, $C_{1,2}$, $C_{3,3}$, and $C_{2,4}$) in the new MKB will result in failure, and hence Bob's device cannot retrieve the new media key. Revocation is effective. Bob's device will not be able to access any content protected with this new MKB. Bob's device will still be able to access content that was protected by an earlier MKB. Thus, revocation of a device is not final; it only depends on the construction of the MKB.

In fact, CPRM uses a far wider MKB than the one used to illustrate this introduction to broadcast encryption. Each device receives 16 device-keys from the Authority managed by the 4C Entity, LLC [276]. The MKB has 16 columns. Its height for DVD-R is 2,500 rows. The MKB is pre-embossed into the lead-in area of recordable media.

To prevent easy bit-to-bit duplication of two recordable media, there is a need to individualize them. For that purpose, each recordable media, in addition to its MKB, which is identical for every instance of media of the same title, has a unique media identifier ID_m. Therefore, CPRM defines a unique K_{mu} that is the result of the hash of the media key K_{media} and the unique media identifier ID_m Thus, $K_{mu} = H(K_{media}, ID_m)$. The unique media identifier is pressed on the media in the burst cut area.[14]

[14] Optical media have two data areas that can be read by a recorder but cannot be written. The first area, the lead-in area, is created when the blank disc is pressed. Thus, it cannot be serialized,

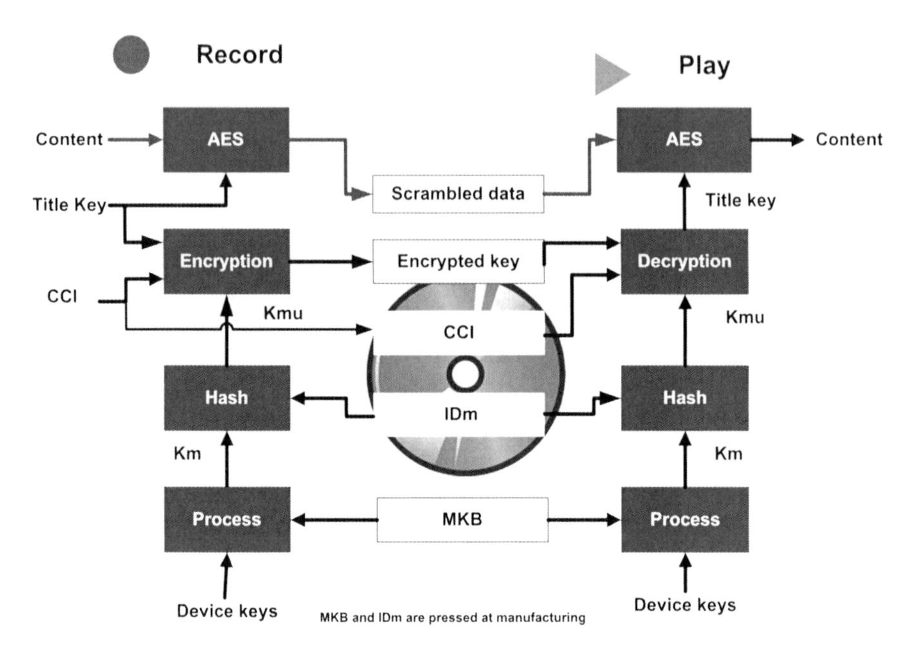

Fig. 8.6 Encryption scheme of CPRM

The content is scrambled using a random key called the title key, K_T. This key is bound to the CCI and the instance of the recordable media with the encrypted key EK_T such as $EK_T = E_{\{K_{mu}\}}(K_T \oplus Hash(CCI))$. The encrypted key is stored together with the scrambled content and the CCI on the recordable media. The CCI defines the usage rights of the recorded piece of content. The use of the hash allows us not to freeze a given syntax for the CCI. Binding the key with the CCI is an elegant way to protect the CCI. An attacker may be interested in tampering with the CCI. For instance, the attacker may overwrite the CCI with the value "copy free", thus totally circumventing the system. DTCP also uses this method (Sect. 9.3).

Figure 8.6 illustrates the recording and playback scheme employed by CPRM. To record a piece of content, the recording device reads from the recordable disc the MKB and the unique identifier ID_m. If the recording device is not revoked, using its device keys, the recorder can extract K_{media} from MKB. Then it calculates

(Footnote 14 continued)
i.e., individualized. The data of the lead-in area are embossed into the glass master. Thus, all the discs from the same glass master will contain the same data in the lead-in area. The second zone is the burst cut area. This area can be written to only with a dedicated laser. Thus, it can be serialized during manufacturing. Each disc can receive a unique serial number, as, for instance, with Sony's ARccOS [259]. This solution increases the price of manufacturing. Of course, no consumer electronics device is equipped with such a laser. Only dedicated professional equipment has such capability.

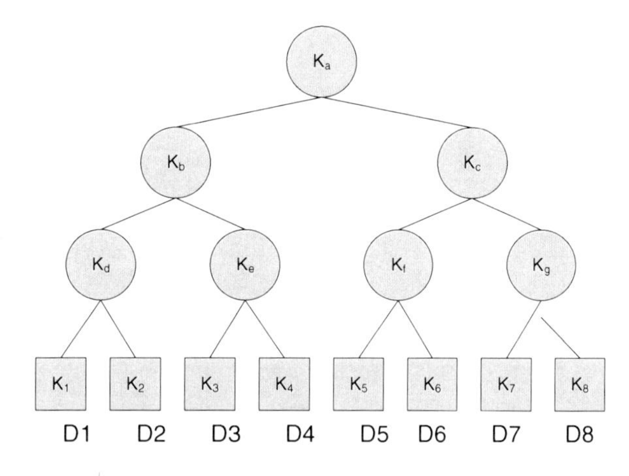

Fig. 8.7 Basic example of broadcast encryption

K_{mu} using K_{media} and ID_m. Then, the recorder generates a random value as title key K_t. It scrambles the data to be recorded with K_t and burns the scrambled data onto the disc. It calculates the encrypted title using K_{mu} and the CCI. It burns both the encrypted title key and the CCI onto the disc.

To playback this content, the player device reads from the recordable disc the MKB and the unique identifier ID_m. If the device is not revoked, using its device keys, the recorder can extract K_{media} from MKB. Then it calculates K_{mu} using K_{media} and ID_m. Then, it reads the encrypted key EK_t and the CCI. If the CCI has not been tampered with, the device can decrypt EK_t using K_{mu} and CCI. With the title key K_t, the device can descramble the stored scrambled data.

One of the main problems with the previous broadcast encryption implementation is that devices share device keys. When a rogue device is revoked, its revocation may affect legitimate devices by invaliding one of their device keys. The larger the number of revocations, the higher the probability of a legitimate device being revoked. Although this probability is extremely low, it is not null. For instance, in the example, the revocation of Bob's device invalidated also one key of Alice's device. If unluckily, three other revoked devices were to use the three remaining keys of Alice, her device would be revoked although she is not guilty. New, better schemes are needed to prevent devices from sharing device-keys. Therefore, new copy protection schemes use more elaborate broadcast encryption schemes based on the notion of trees. Depending on the scheme, the type of tree may differ.

For this pedagogic example, we will simplify the problem by using a typical binary tree and we will manage a maximum of eight devices $\{D1...D8\}$. Each leaf of the tree represents one device.

An Authority manages broadcast encryption. This Authority generates a random key for each node of the tree $\{K_a, \ldots, K_g\}$. The key at the root of the tree is K_a. The Authority generates a random key for each leaf of the tree $\{K_1, \ldots, K_8\}$. K_i is associated with device D_i. Figure 8.7 illustrates this key distribution. Then,

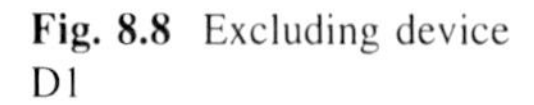

Fig. 8.8 Excluding device D1

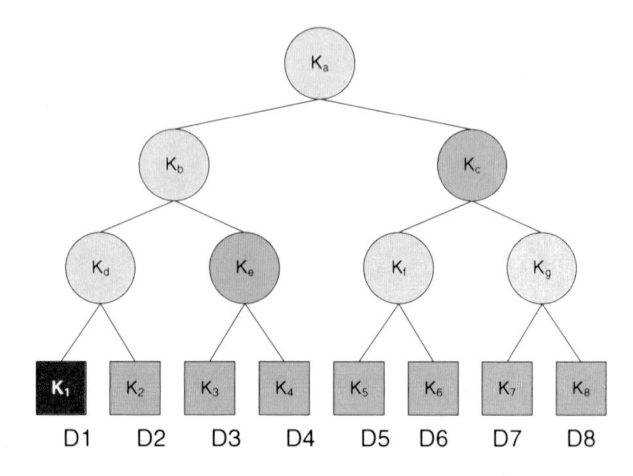

the Authority dispatches these keys to the devices. A device receives all the keys whose nodes are on the path to its leaf. Thus, device D_1 receives the set of four keys $\{K_a, K_b, K_d, K_1\}$ whereas device D_6 receives the set of four keys $\{K_a, K_c, K_f, K_6\}$. These sets of keys are called the device keys. The assumption is that the device keys are securely stored in the device. They should be protected in terms of confidentiality and integrity.

Once the device keys are distributed to the manufacturer, it is possible to address any subset of devices. Let us suppose that we want to securely transfer message M to all the devices. We generate a random key K_M. We encrypt M with the symmetric encryption algorithm E. The ciphertext is $M\prime = E_{\{K_M\}}(M)$. The Authority generates the corresponding MKB. For that purpose, it selects the nodes such that all the authorized devices are in the paths defined by these nodes, whereas the unauthorized devices are not. It encrypts the random key K_M with another algorithm E' using the keys of the selected nodes.

As we want to address all the devices, the Authority selects K_a and generates $MKB1 = \{E\prime_{\{K_a\}}(K_M)\}$. We can now distribute M' and MKB. Every device has K_a in its set of device keys; thus it can decrypt MKB1 and retrieve K_M. With K_M, it can decrypt M' and thus access message M.

Let us now address all the devices except D1. In this case, the Authority selects K_C, K_e, K_2 (Fig. 8.8). It generates the MKB2 by encrypting $K_M$with each of these keys. $MKB2 = \{E\prime_{\{K_c\}}(K_M), E\prime_{\{K_e\}}(K_M), E\prime_{\{K_2\}}(K_M)\}$. We distribute M' and MKB2.

When device D_1 receives MKB2, it attempts to decrypt MKB2 with each of its device keys, i.e., $\{K_a, K_b, K_d, K_1\}$. None of these keys will be able to retrieve K_M. For all the other devices, there will be one of the device keys that will be able to retrieve K_M. For instance, device D_6 has the device keys $\{K_a, K_c, K_f, K_6\}$; thus when using K_c, it will be able to retrieve the first value of MKB2.

The size of MKB increases with the number of revoked devices. Research in broadcast encryption explores more complex schemes of trees than the binary one that we used to illustrate the concept [277]. In fact, no deployed system uses such a rudimentary tree structure.

8.5 Blu-Ray Discs

8.5.1 AACS

With the advent of two new formats for high definition discs, HD DVD and Blu-ray, a set of companies decided to design a new copy protection scheme for HD-DVD. In 2004, Disney, Intel, Microsoft, Panasonic, Warner Bros., IBM, Toshiba, and Sony formed the AACS group. In 2006, they issued the first specification of the Advanced Access Control System (AACS). They founded the AACS Licensing Authority (AACS-LA), in charge to handle the compliance and robustness regime, as well as secret data. Although HD-DVD died, the Blu-ray consortium endorsed AACS with an additional optional security layer called BD+.

AACS specifies many security features [278]:

- encryption,
- sequence keys, and
- watermarks

The architecture of encryption of AACS is rather similar to those of CSS and CPRM. Figure 8.9 describes the creation and the playback processes of an AACS-protected disc. There are three stages of encryption and keys. AACS uses AES with 128-bit keys.

The replicator, i.e., the manufacturer who presses the discs, generates for each piece of new content a random 128-bit title key. The piece of content is AES-encrypted with the title key. The title key is encrypted using the volume unique key K_{vu}. The result is the encrypted key. The volume unique key is the result of a cryptographic hash using the Volume ID, the usage rights defined for the disc, and the media key K_m. The Volume ID is data defined by the replicator. As for CPRM, its goal is to thwart genuine bit-to-bit copy. The Volume ID is located in a BD-ROM Mask, the equivalent of the DVD's lead-in area, i.e., a zone that DVD/Blu-ray recorders cannot overwrite. Furthermore, only an AACS-compliant drive can read this zone. Indeed, only licensed Blu-ray disc manufacturers have the equipment to write to this area. AACS-LA provides the secret media key to the replicator. A MKB protects the media key K_m (in a way similar to that described in Sect. 8.4[15] although it allows revoking a device despite this affecting other devices). AACS-LA delivers the MKB

[15] AACS uses a variant of broadcast encryption called subset difference [277].

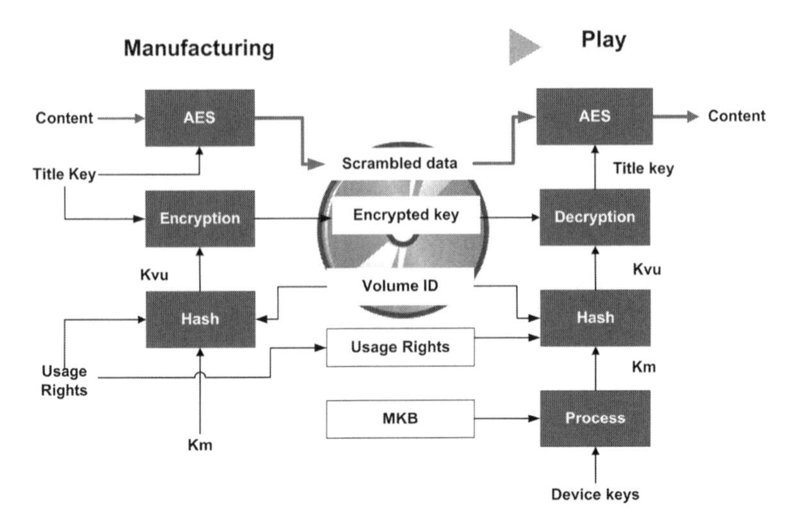

Fig. 8.9 Scrambling of AACS

together with the corresponding media key. Scrambled data, encrypted key, usage rights, volume ID, and the MKB are pressed on the Blu-ray glass master. The glass master stamps the final Blu-ray discs.

At playback, the MKB provides the right media key K_m if the device is not revoked. The hash of the media key, the Volume ID, and the usage rights are used to retrieve the volume unique key K_{vu}. The volume unique key allows decrypting the encrypted key and thus provides the title key. The title key allows decryption of scrambled data and thus retrieves the content in the clear.

The Devil's in the Details

The first attack on the AACS taught us an interesting lesson [185]. In January 2007, a few months after the launching of AACS, a hacker by the name of muslix64 posted on the Doom9 forum a piece of software, `HDDVDbackup`, which emulated AACS decryption. The software needed the secret title keys that protected disc. It was an implementation of the publicly available specifications.

A few days later, crackers published the first title keys. The extraction method was extremely simple. The 128-bit AES title key protects AACS content. The crackers explored 16 consecutive bytes of the RAM used by a commercial AACS-compliant software-based player. For each stored value, they attempted to decrypt the scrambled content. The memory space to explore was reduced from the initial 2^{128} to less than 2^{30} (the size of the used RAM), although the key still had its entropy of 2^{128} bits. Suddenly, brute force was feasible. This initial software implementation of AACS was too simplistic. It was not a secure implementation. The attackers extracted the keys without having knowledge of the actual implementation. They did not have to reverse

engineer the code of the software. A few weeks later, another cracker, by the nickname of Arnezami, managed to extract the volume unique key. Another few weeks later, another cracker, by the nickname of Atari Vampire, managed to extract the device key of the same software [279]. Once more, the secret key was located on consecutive bytes, near the location spotted by Arnezami. Of course, the implementation of the commercial package was fixed, and the keys were properly obfuscated in a new version.
Lesson: Handling secret keys using software is an extremely important and difficult task. Handling must be as stealthy as possible. Security designers follow many rules, such as never transferring a key in one block, since it may be observed on the data bus; never storing keys in the same location; never storing keys in contiguous memories, but spreading them; and always erasing the temporary locations of keys once they have been used.

8.5.2 Sequence Keys

As illustrated, AACS implements an efficient revocation mechanism for a set of devices. Nevertheless, revocation relies on an initial assumption: the authority knows the identity of the forged or badly behaving devices. Thus, it needs the ability to trace the player that would have leaked a piece of content on the Internet. This is called traitor tracing (Sect. 3.3.3). The idea is to individualize (or serialize) each actual rendered content. This is rather simple in the case of streaming, especially when serializing occurs at the head-end side. The challenge is more difficult in the case of pre-recorded content. All discs are identical.

AACS uses a technique called sequence keys. The idea behind it is to select some short sections of the movie, each of about 5 s. Each section is slightly modified to create different versions of the same section [280]. When playing back the movie, each player may use a different set of these individualized sections. Figure 8.10 gives an example with two individualized sectors. Each sector has five variations. One player may render the sequence {*section* 1, *section* 2.2, *section* 3, *section* 4.5, *section* 5} whereas another player may render {*section* 1, *section* 2.3, *section* 3, *section* 4.1, *section* 5}. If the viewed path is deterministically defined by the identity of the viewer, then it should allow tracing the player. Obviously, in the case of pre-recorded content, each variation increases the size of the disc, which may be an issue. In the case of AACS, the traitor-tracing scheme has three requirements:

- it must not increase the final size of the disc by more than 10 %;
- it must be able to support a huge number of devices; and
- the number of movies needed to successfully trace a player should be minimal.

In the case of AACS, the traitor-tracing scheme uses two nested, coding schemes. The first scheme, called "inner code", is movie-dependent. Each movie

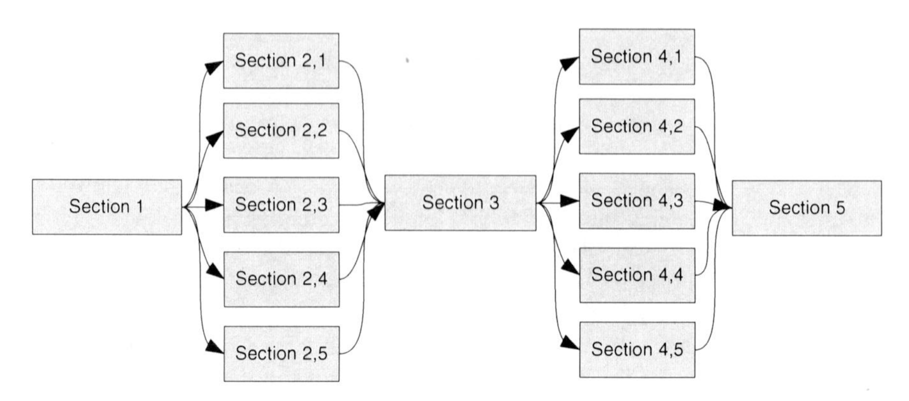

Fig. 8.10 Traitor tracing by sequence keys

has 16 dedicated sections.[16] Each section exists with 16 variations. The variation is done by using a watermark with a different payload for each variation. In other words, each movie generates 256 different versions. The second scheme, called "outer code", attributes to a player a set of movies, thus allowing an increase of the size of the actual code. The "outer code" is device-dependent. Each device receives a set of 256 sequence keys. No two devices have the same set of device keys. Each sequence key descrambles one of the 16 variations of a given section, i.e., enforces the "inner code".

There is no public evidence that sequence keys are actually implemented in current Blu-ray discs.

8.5.3 BD

The Blu-ray association specified an additional layer of security, called BD+, which sits on top of AACS. The purpose of this layer is to recover from potential security breaches occurring in the AACS. BD discs may contain a title-specific piece of software, called the applet, that executes on a BD player before and during the movie playback. The program is written in a dedicated interpreted language: Self-protecting Digital Content (SPDC) [281]. Figure 8.11 describes its main elements. SPDC is mainly a Virtual Machine and a Media Transform block. The Virtual Machine is far simpler than Java, another well-known Virtual Machine. It is similar to the assembly language of early microprocessors.[17] The Media

[16] The numerical values of the parameters such as the 16 sections or the 256 versions are not confirmed. They are deduced from IBM's published work, which served as the foundation for AACS's sequence key [280]. The real values are not public.

[17] Readers familiar with old microprocessors 6502 or 8051 would feel at home with the SPDC virtual machine.

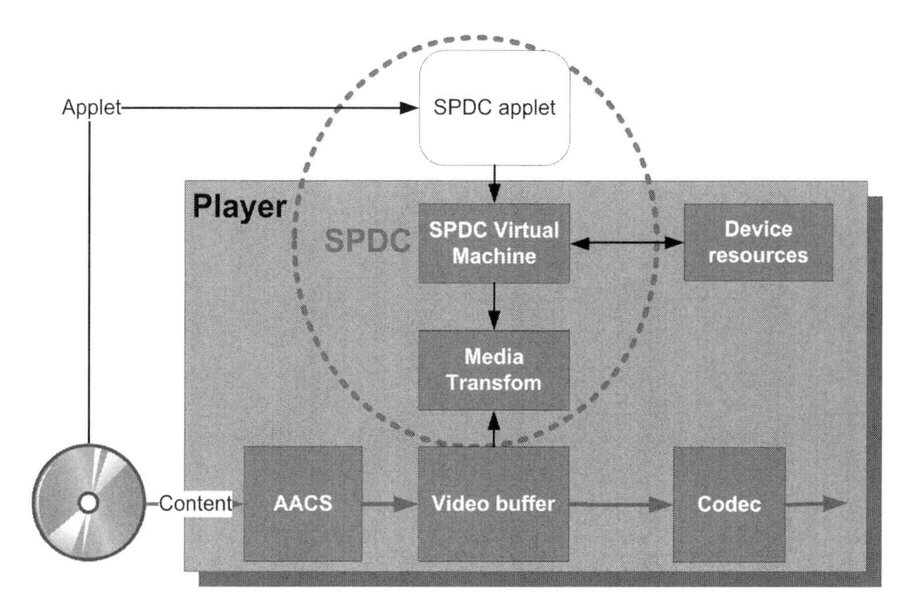

Fig. 8.11 Basic synoptic of SPDC

Transform block can apply basic modifications to the compressed essence. When reading a BD+ disc, the player loads the applet. Then the player executes the applet. Usually, it first checks whether the host player is compliant and not revoked. Then, the applet provides configuration parameters to the Media Transform block. This block deterministically modifies, using the parameters, the compressed content, for instance, swapping some bytes in the stream.

In the BD environment, the applet is stored on the disc. This applet modifies the AACS-decrypted content before video decoding. There are mainly two potential usages:

- Renewability of the security of BD: If ever AACS were to be cracked, the use of an applet would allow us to heal the broken AACS. At the authoring stage, the essence is modified with a function f before being AACS-encrypted. Thus, on the disc, there is $AACS(f(essence))$. The applet that defines the reverse transformation, i.e., f^{-1} is delivered within the Blu-ray disc. When the disc played back, the applet would apply f^{-1} to the AACS decrypted essence, i.e., $f(essence)$. It is assumed that non-compliant players would not properly execute this applet or that the applet would detect non-compliant players. Thus, if the applet does not execute, the non-compliant player will try to render $f(essence)$ rather than the essence itself; thus, the rendering would fail. Ideally, an applet should be unique for each title. This limits the risk of the reuse of a past exploit. The crackers should have to reverse the applet for each new title, increasing the required investment for pirates. Obviously, this would increase the cost of "authoring" the applet and manufacturing the disc.

- Forensics: The applet should provide parameters to the Media Transform block so that each piece of content is unique for its player. In other words, the Media Transform block embeds a watermark. This watermark is different from the watermark defined by the sequence keys. Thus, it should be possible to trace the player that issued one piece of content. SPDC can collect information and access the internal memory of the players (peek and poke instructions). The collection of these data may uniquely fingerprint the infringing player. The applet builds a unique payload derived from the collected information. The extraction of the watermark from illegal content points the incriminated player. Through the revocation of MKB, it is possible to annihilate the infringing players.

For this usage, the basic assumption is that it is possible to perform watermark embedding through basic media transforms. This is possible with a specific type of watermarking process: the two-step watermark [282]. The content to be watermarked is first preprocessed offline. All the heavy digital signal processing occurs during this step. The result is a set of possible positions within the compressed information that can be modified while the result remains invisible. With each position is associated a set of potential replacement values. The embedding process occurs in real time. It consists of parsing the content to watermark, locating the positions that the preprocessing identified, and, depending on the expected payload, transforming or not transforming the corresponding data. This embedding is extremely cost-efficient in terms of calculations. The underlying watermarking technology used by the preprocessing uses the typical techniques described in Sect. 3.3.1.

8.5.4 Watermarks

AACS discs can potentially use three types of watermarks:

Sequence keys, as described in Sect. 8.5.2, use video watermarks to implement traitor tracing. The watermarks are embedded at the authoring stage. The watermarking technology is not standardized.
SPDC applets can implement video watermark functions. Their goal may be to implement traitor tracing or to embed additional information that may complement the traitor-tracing scheme of sequence keys. They may, for instance, embed data describing the configuration of the leaking player. The watermark is embedded at playback. The watermark technology is not standardized.
The Cinavia audio watermarking system signals special content, such as professional content. Cinavia is developed by Verance [283]. The AACS standard makes its implementation mandatory for any Blu-ray player [284]. The watermark is embedded at the authoring stage. Cinavia's watermark encodes information indicating specific usage rights:

- The piece of content is limited to professional usage. Regular consumer electronics devices should not play it back. This can be used for signaling screeners or theatrical releases. Theater-captured content distributed with a Blu-ray pirate disc should still carry the audio watermark. A Blu-ray player should terminate playback upon detecting the mark.
- Only professional equipment, such as the one used by professional replicator, should allow copying of this piece of content. Regular consumer electronics recorder should refuse to copy it upon detecting this watermark.
- Consumer electronics players that implement the Cinavia detector should trigger the potential presence of an audio watermark and obey correspondingly.

Chapter 9
Protection Within the Home

9.1 The Home Network Problem

The digital home network handles a plurality of devices using different media such as Ethernet, General Packet Radio Service, Wi-Fi, and USB. The home network encompasses fixed devices as well as mobile devices. It encompasses consumer electronics (CE) devices as well as personal computers and mobile phones (see Fig. 5.1, Delivering video to the home).

In a simplified model of the home network (derived from DVB-CPCM [173]), there are four roles:

- The acquisition point is the location where a piece of content enters the home network. The acquisition point may receive content through broadcast, broadband, or portable storage media. A home network may have several acquisition points: a gateway that receives broadband and broadcast services, a mobile phone, or a DVD reader. Often the ingested content is protected by CAS or DRM technology.
- Storage points store the content in the home network. We may consider different types of storage media: hard drives, optical recordable discs, and flash drives. In the case of hard drives, the actual storage may be an embedded hard drive (as in a Personal Video Recorder (PVR)), inside the local area network, or even somewhere in the Internet cloud.
- The rendering point is the place where the content is rendered for use by the consumers. It is often a TV set, but may also be the video card of a computer, or the display of a mobile device.
- Exporting points transfer pieces of content out of the home network. The content is always in compressed mode.

A device may simultaneously play several roles.

DRM/CAS protect content until it reaches the acquisition point. The location of the acquisition point within the home network is a tricky issue known as CA-to-DRM (Sect. 9.2). Content transferred from the acquisition point to other points

E. Diehl, *Securing Digital Video*, DOI: 10.1007/978-3-642-17345-5_9,

needs protection. The first approach is to use link encryption. This is usually done by DTCP (Sect. 9.3). DVB-CPCM proposes a more global approach (Sect. 9.5). Content always stays scrambled within the home network. Descrambling only occurs at rendering and exporting points.

9.2 Termination of DRM/CAS

Copyrighted content may be protected when delivered to the home network, for instance, through DRM for Video On Demand, CAS for broadcast programs, or AACS for Blu-ray discs. As the differences between CAS and DRM have tended to vanish, we will use DRM as a generic term in the rest of this section. It encompasses DRM, CAS, and protection of prerecorded or recordable media.

With the advent of the home network, DRM has had to hand over content to other copy protection schemes that will take care of it on the home network. This problem is often called the CA-to-DRM (CA2DRM) problem. One of the most crucial issues is where to end the DRM protection. In the security architecture of home networks, there are two zones. The first zone is protected by the DRM of the content providers. The second zone is accessible to content once the delivering DRM has granted access. Obviously, there is a need for content protection in the second zone. There are mainly two types of solution.

In Translating acquisition, the acquisition point hosts the delivery DRM. If the user is entitled to receive the content, the delivering DRM will translate it into a common content protection system. The rendering points host this common protection system (CPS).[1] Thus, if a rendering point supports the media format, it can, in principle, render any content received by any acquisition point. Fig. 9.1 illustrates this architecture.

This solution offers an elegant solution to the interoperability of DRMs. Interoperability occurs at acquisition. Often, the acquisition point is dedicated to a content provider, as is the case for STBs. Thus, it can easily be specialized for the corresponding DRM solution. The translation into common CPS provides the required interoperability solution. This is only possible if the DRM provider accepts handing over content to the common CPS.

A basic example of implementation of this scenario is when content is stored at the acquisition point (typically on a PVR) under the delivery system protection. Upon consumption, content is transferred to the TV set through a link protection system (e.g., DTCP).

Through *traversing acquisition*, the acquisition point acts as a pass-through for the DRM. The rendering point hosts the delivery DRM. Interoperability requires either that the rendering point hosts all possible DRM solutions (as illustrated

[1] A CPS is a generic term used here for content protection systems that do not deal with the content delivery. It includes PVR storage protection, link protection (DTCP-IP), or any other system local to the home networks. It may also sometimes be seen as a local DRM.

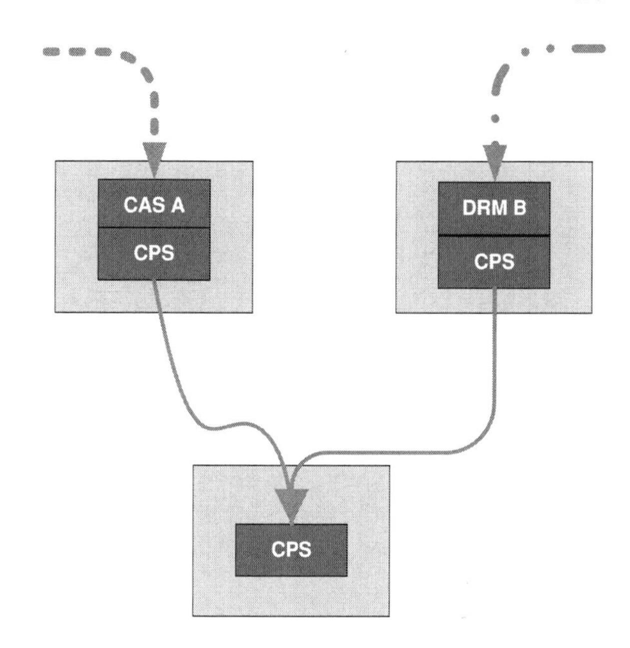

Fig. 9.1 Translating acquisition strategy

in Fig. 9.2), or that a more complex system require the translation of license or content by a remote server (as illustrated in Fig. 9.3).

The first version is implementable with difficulty if there are too many systems to support. Furthermore, it seriously impairs the cost of the rendering point.

In the last traversing solution (Fig. 9.3), the rendering point hosts one single DRM. Thus, it requires the connection to an online server specific to the sink DRM system (DRM A here). The acquisition point with DRM B forwards content protected by DRM B to the rendering point. The rendering point is not able to unprotect this content. It sends information about this content with a request for translation to the online server. The online server checks within the digital rights locker whether it is authorized to perform the translation. If this is the case, it returns the license for the content, but for DRM A. The digital rights locker is a repository of the rights acquired by the customer for her content. Each DRM participating in the system records information on each content purchase of each consumer and the associated usage rights. That way, when the DRM A server connects to the digital rights locker, the digital rights locker may validate the request from a consumer who has already purchased the content and verify whether the associated usage rights authorize the transfer of the content to the rendering point.

This latter approach can be further refined into two sub-approaches:

- The online server only manages usage rights and keys: when transferring content from DRM A to DRM B, a new set of usage rights and descrambling keys shall be obtained from the digital locker, the content itself being transferred

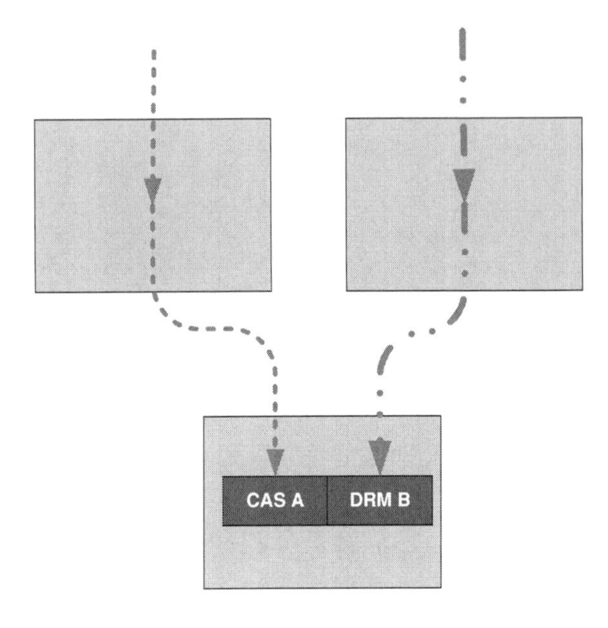

Fig. 9.2 Traversing acquisition strategy (no interoperability)

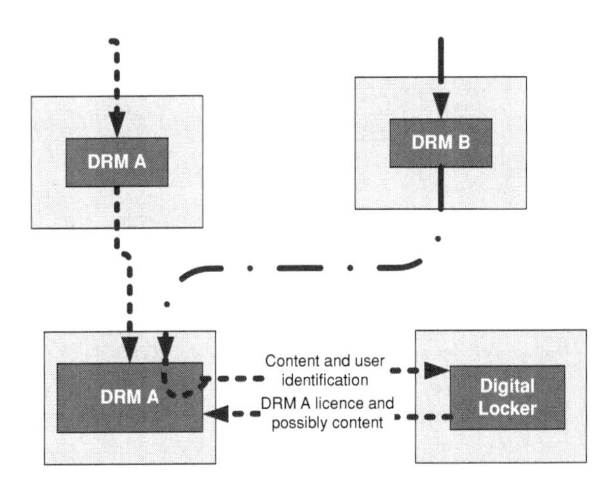

Fig. 9.3 Traversing acquisition strategy (with interoperability)

locally (this requires a standardized format and encryption for the essence of the content). The usage rights and keys are packaged within a license.

- The online server manages usage rights and content: in this case, content shall practically be reacquired each time it is to be transferred from one DRM to another (which may require huge broadband capability at the server side). In other words, the online server delivers the same piece of content but protected by DRM A rather than DRM B.

The first approach has the advantage of being simple and not requiring a lot of bandwidth. Unfortunately, it requires the standardization of a content format. This is a difficult commercial, political, and industrial challenge (Sect. 11.2.5, Interoperable Rights Locker). The second approach requires less goodwill of the stakeholders. Nevertheless, it puts a heavy burden on the bandwidth of the network. Thus, its operational cost is far higher than the first approach.

9.3 DTCP

In 1996, the CPTWG created the Digital Transmission Discussion Group. Its call for proposals received 11 submissions. In 1998, under the mandate of CPTWG, Hitachi, Intel, Matsushita, Sony, and Toshiba, the so-called "5C", designed the first version of DTCP. The technology is licensed by the Digital Transmission Licensing Administrator, (DTLA) LLC. DTCP cryptographically protects audio and video content while it traverses digital transmission mechanisms [285]. It was originally designed to support IEEE 1394 (also known as Firewire).[2] Meanwhile, DTCP has been extended to support Ethernet IP (in which case, it is called DTCP-IP), USB, Opt-i Link, and MOST. Designed by Sharp and Sony, Opt-I Link is an extension to IEEE1394-2000 using optical fiber. MOST, the Media Oriented Systems Transport, is a digital bus for media and entertainment in the automotive market.

DTCP handles four copy usage rights:

- Copy Never: The sink device should not make any copy of content marked "copy never".
- Copy Once: The sink device is authorized to make one copy of content marked "copy once". This copy is then marked "Copy No More". In fact, copy once is copy only a first generation. Indeed, it is possible to make sequentially an unlimited number of copies of "copy once" content.
- Copy No More: The sink device does not make any copy of content marked "copy no more". This mode in fact enforces "copy once". "Copy no more" content is the result of the duplication of "copy once" content.
- Copy Freely: The sink device has no restriction in making copies.
- In the case of DTCP over IP, there is an additional flag, which indicates if the content should remain local. Local content can only be carried in a Local Area Network, and not in a Wide Area Network. This restriction is enforced by restricting the number of traversed hops to three (using Time To Live (TTL)[3]

---

[2] In the late 1990 s, it was foreseen that the preferred bus for home networks would be IEEE1394. By default, IEEE1394 features quality of service (QoS) which is paramount for digital video transmission. This was not the case with Ethernet IP. Slow adoption of IEEE1394 allowed IP to solve the issue and propose QoS. IP is now the undisputed technology of choice for home networks.

[3] In IP protocol, it is possible to define in the header the maximum number of hops, i.e., the number of traversed network devices. Once this value is reached, the corresponding IP packet is automatically discarded.

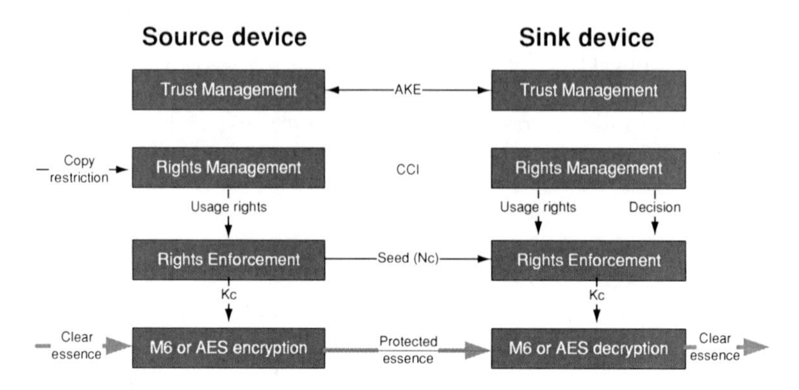

Fig. 9.4 Four-layer model of DTCP

parameter) and by setting the measurement of the Round-Trip Time (RTT)[4] to less than 7 ms.

These four states are summarized in one two-bit indicator called the CCI.

Within DTCP, content flows from one source device to a sink device (Fig. 9.4). When a source device sends content to a sink device, the first step is the Authenticated Key Exchange (AKE). It starts by mutual authentication between source and sink devices. DTCP supports two types of authentication: the full authentication for devices that support all the copy restriction modes, i.e., the full range of CCI, and the restricted authentication for devices supporting fewer features (in fact, only copy never and copy free). Full authentication uses Elliptic Curve Digital Signature Algorithm (EC-DSA) and Elliptic Curve Diffie–Hellman (EC-DH) in a typical challenge-response protocol. Each device has its own certified key pair. DTLA delivers the keys together with the signed certificate. The restricted authentication is for less powerful devices with fewer capabilities. Each device has a set of 12 keys. DTLA delivers these keys to the manufacturers. Through an answer to a random challenge, a device proves the challenger that it holds a set of keys delivered by DTLA. Both authentication schemes establish a common exchange key. Using two types of authentication was a simple method to reduce the cost. Restricted authentication requires less computing power than full authentication, and thus does not put unnecessary burden on cheap devices.

Once the source and sink devices have been mutually authenticated, the source sends scrambled content to the sink. For that purpose, it can use either M6 [286] or 128-bit AES. M6 is a proprietary algorithm designed by Hitachi. For DTCP-IP, the scrambling algorithm is 128-bit AES. The source device defines the content key K_c used to scramble/descramble the piece of content. It is the result of a function using the common exchange key defined during the AKE phase, a random seed

[4] RTT is a measure of the current transmission delay in a network, found by timing a packet bounced off some remote host. This can be done with the ping command. The restriction uses an average value calculated after several trials.

generated by the source, and a value depending on the value of the CCI. The CCI is embedded in the clear in the content stream. The source device sends the seed to the sink device.

The rights enforcement layer of the sink device combines the received seed, the received CCI, and the exchange key to recover the content key K_c. If no data has been impaired, the sink can calculate the proper value of key K_c and decrypt the content. CCI information is forwarded to the device for a decision what to do with the descrambled content. For instance, a recorder will refuse to record Copy Never and No More Copies content.

DTCP defines a typical revocation mechanism using System Renewability Messages (SRM). An SRM is a CRL carrying the list of revoked devices. The content delivers the CRL to the device. When a DTCP device receives an SRM, it checks if the message is more recent than its current version. If this is the case, the device updates its list of revoked devices. During the authentication phase, the device checks if its interlocutor is on the revocation list. If so, the authentication fails. DTLA generates the new revocation lists and signs them using EC-DSA.

9.4 HDCP

HDCP was designed by Intel and Silicon Image. The first specification was published in February 2000 [287]. Its primary goal was to protect the Digital Video Interface (DVI) and HDMI buses. It is the de facto standard for these interfaces. Since 2000, it has been extended to many other interfaces such as DisplayPort and Wireless High Definition Interface (WHDI). It is licensed by Digital Content Protection LLC.

As indicated in Fig. 9.5, authentication is the first step. A HDCP transmitter checks whether the receiver is authorized to receive HDCP-protected content. According to a principle similar to that of broadcast encryption (Sect. 8.4, CPPM CPRM), each entity has a table of 40 private keys (noted *KSV*) and a single identifier. Each *KSV* key is 40 bits long. Furthermore, each key has the same number of bits set to "1" and "0". Transmitter A sends its identifier A_{KSV}. Receiver B returns its identifier B_{KSV}. Transmitter A checks that B_{KSV} is not revoked. The receiver must have the most recent revocation list. Now the transmitter A knows that receiver B is a compliant HDCP receiver. They establish a common session key K_m. Each device uses its table KSV and the other device's identifier to calculate K_m. Thus, each device can find K_m without exchanging it over the bus. Only the devices having a valid table can establish a common key. Receiver B encrypts a random value R_0 with key K_m. Receiver B sends R_0 and the computed value to transmitter A, which checks whether they match. Transmitter A has proof that receiver B has been authenticated and that they share the same key. It should be noted that due to the one-way transfer of content (from transmitter to receiver), only the transmitter needs to authenticate, but both sides have to be HDCP-compliant. A compliant transmitter does not want to send any protected

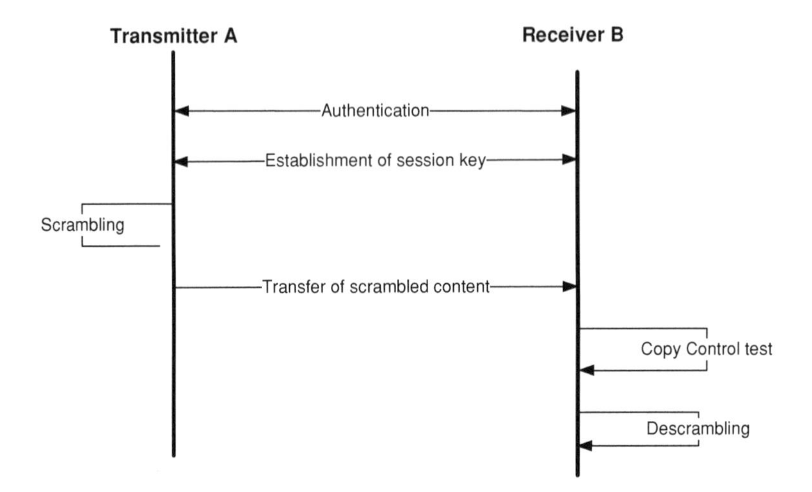

Fig. 9.5 HDCP key management

content to a rogue receiver. A compliant receiver does not care if it receives content from a compliant transmitter or rogue one.

The encrypted transmission can begin. Scrambling is a basic stream cipher using a 56-bit key. The content is bitwise XORed with the output of a pseudo-random generator initiated with the key K_m. Receiver B performs the same operation, i.e., a bitwise XORed with the output of the same pseudo-random generator. In the case of HDCP, there is no test of copy conditions because the compliance rules enforce that the receiver cannot replicate received content. In other words, the content is mandatorily "copy never".

HDCP is widely deployed across the world as the protection mechanism for HDMI. Nevertheless, some competitors have appeared. In China, the Unified Content Protection System (UCPS) is expected to replace HDCP in the local market [288]. It is in the draft stage for comments.

The security of HDCP is weak. The size of the keys (respectively 40 and 56 bits) is far too small compared to current best practices [71]. Even at the time of design, AES had already been selected, and the recommended key size was greater than 90 bits. Furthermore, the actual key space is reduced to 37 bits rather than 40 bits (due to the parity of the bits). Thus, not surprisingly, the system was attacked. Once more, as with CSS (Sect. 8.2), we may assume that the choice of these small keys was driven by the needs of being early in the market. At this time, legal restrictions of cryptographic algorithms required short keys.

In 2001, cryptographers published the first attack [289]. They analyzed the key distribution scheme used by HDCP. They proved that for the system to work, every *KSV* was derived from one unique master key. Furthermore, they demonstrated that analyzing the secret keys of 40 devices would be enough to find the master key. Knowing the master key, it would be easy to calculate any private key using the public key. Meanwhile, another researcher, Keith Irwin, published four basic attacks to defeat HDCP [290]. In September 2010, an anonymous cracker published the master key [291]. There is no more a way to revoke any device.[5]

9.5 DVB-CPCM

The third interesting initiative is DVB Content Protection and Copy Management (DVB-CPCM) [292]. CPCM was designed by the DVB Copy Protection Technical working group and standardized in 2009 by ETSI under reference TS102 825. The DVB project gathers about 250 companies involved in the video broadcasting business, including service operators, regulators, consumer electronics manufacturers, and conditional access providers. CPCM itself was agreed upon by consensus in a technical working group involving major industry players (including Sony, Panasonic, Thomson, NDS, Nagravision, Irdeto, Sky, BBC, EBU, IRT, Disney, Warner, and MPA). DVB-CPCM is currently the most sophisticated content protection system for home network.

Starting around 2000, the design of DVB-CPCM took many years. The first step defined a set of agreed-upon commercial requirements. The objective was not to design simple copy protection but a complete content management system that would embrace the complexity of future home networks. The goal was to have pure copy protection technology that would protect content handed over by CAS or DRM without lowering the overall security. The following is a non-exhaustive list of these requirements.

- DVB-CPCM must allow content to be exchanged in a user domain including many devices. These devices may be fixed, remote, or portable. The devices may not all be persistently connected to the local area network. The devices may be working in untethered mode.
- DVB-CPCM must restrict its content protection layer either to the essence or to the ISO transport layer.
- DVB-CPCM must define a CA2DRM (Sect. 9.2) flexible enough to allow proprietary security levels and enable countermeasures.
- The CA2DRM specifications must not lower the security of CAS or DRM.

DVB-CPCM is content-centric and not device-centric. A piece of DVB-CPCM protected content is associated with a content license. The content license carries the usage rights and the keys needed to descramble the content.

DVB-CPCM defines a set of functional entities. Each functional entity is an abstraction of the different types of content management. A device may encompass several functional entities. DVB-CPCM defines the five functional entities depicted in Fig. 9.6.

---

[5] The revocation of HDCP has never been activated, although many HDCP strippers, i.e., gimmicks that removed HDCP protection, are available on the market.

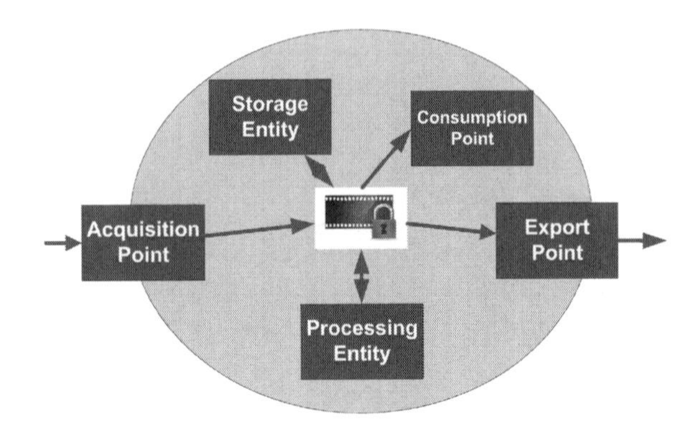

Fig. 9.6 DVB-CPCM functional entities

- The Acquisition Point receives content protected by CAS or DRM, or even in the clear from Free to Air. Within the acquisition point, the CAS or DRM hand over the content to DVB-CPCM. DVB-CPCM's content protection replaces the CAS/DRM content protection. The input CAS/DRM usage rights are translated into DVB-CPCM usage rights.
- The Processing Entity performs transformations on the content such, as transcoding and image handling. Once processed, the content remains protected by DVB-CPCM with the same usage rights.
- The Consumption Point renders the content to a display according to the usage rights defined in the content license.
- The Export Point has the reverse function of the acquisition point. The DVB-CPCM content is handed over to other content protection systems. The current usage rights are translated into the usage rights of the exported content protection system.
- The Storage Entity just stores the DVB-CPCM content.

DVB-CPCM assumes that all devices are connected through a two-way communication network. This link does not need to be permanent, thus supporting untethered mode. DVB-CPCM relies on other standards to perform the actual communication. It may, for instance, use the DLNA standard [293].

One interesting concept of DVB-CPCM is the Authorized Domain. The Authorized Domain encompasses all the devices belonging to a family. A device belongs to one Authorized Domain. DVB-CPCM content is bound to an Authorized Domain, thus being accessible to any device belonging to the Authorized Domain.

The management of the Authorized Domain mainly oversees a device joining an Authorized Domain or leaving an Authorized Domain; hence, the management is simpler. Three parameters define the size of the Authorized Domain.

- The ceiling value that defines the maximum number of devices that can simultaneously belong to the same Authorized Domain.

- The ceiling value that defines the maximum number of remote devices that can simultaneously belong to the same Authorized Domain. Remote devices are connected through a WAN.
- The ceiling value that defines the maximum number of local devices that can simultaneously belong to the same Authorized Domain. Local devices are connected through a LAN.

The DVB-CPCM specifications do not define these ceiling values. They will be defined by the C&R regime. As long as the ceiling value is not reached, any device can join the Authorized Domain. The management of the Authorized Domain is autonomous and does not require any connection to a remote central server (as, for instance, in OMA or Coral). As the devices of an Authorized Domain can be portable or remote, DVB-CPCM is an elegant solution for interoperability.

DVB-CPCM supports a rich set of usage rights. This rich expressivity was necessary to support future business models envisaged by CAS and DRM technology providers. The study of these usage rights illustrates many scenarios of the use of DVB-CPCM. Following is a non-exhaustive list of the supported types of usage rights.

- *Copy Control Information* defines the typical conditions that allow or do not allow copying of content. In addition to the typical four states of CCI, DVB-CPCM defines also "Copy Control Not Asserted", "Copy Once", "Copy No More", and "Copy Never".
- *Viewable flag* defines if a piece of content can be viewed within an Authorized Domain. It is thus possible to state that a piece of content will just transit through the Authorized Domain before being exported to other copy protection schemes. Another scenario is "push VOD". The piece of content is pushed inside the Authorized Domain to the storage entity independently of any consumer action with its viewable flag disabling viewing. Once the consumer purchases this content, the DRM will correspondingly set the viewable flag.
- *View window or period* defines how long a piece of content can be viewed within the Authorized Domain. The period is defined either as absolute time or as duration after the first playback.
- *Simultaneous View Count* defines how many consumption points of the Authorized Domain can simultaneously access the same piece of live content.
- *Propagation information* restricts the actual transfer of content. The authorized transfer may be any of the following:
  - *Local*, in which case access to the piece of content is restricted to the subset of devices of the Authorized Domain belonging to the same Local Area Network. The restriction is enforced through the definition of the maximum number of traversed hops (using the Secure Time To Live (STTL) parameter) and the measurement of the Round-Trip Time (using SRTT). As regular TTL and RTT are not protected, DVB-CPCM has developed a new flavor with the same characteristics but one more secure, i.e., its parameters cannot be impaired and returned values are trusted.

  - *Geographically limited*, in which case the piece of content is restricted to the subset of devices of the Authorized Domain that verifies the geographical constraints. This is only possible with devices that have explicit knowledge of their geographical location. DVB CPCM proposes using a set of geo data methods such as IP address locations and country code. If the device does not have this optional feature, the constraint is restricted to a localized Authorized Domain.
  - *Full domain*, in which case the piece of content can be transferred in any part of the Authorized Domain, even through a Wide Area Network.

- *Remote access dates and flags* can temporarily restrict the propagation restrictions. This is useful, for instance, in case of blackout zones. In that case, the enforcement is limited in time. After a while, the content can be transferred all over the Authorized Domain.
- *Analog export or output controls* define the usual output controls such as those for downsizing or activating Macrovision's APG.
- *Export to other copy protection technologies* identifies to which known technologies it is acceptable to export and under what conditions.

Within the DVB-CPCM Authorized Domain, content is protected using the Local Scrambling Algorithm based on AES 128-bit with CBC or RCBC mode. The Content License carries the AES key together with the usage rights. The Content License is signed to ensure its integrity. Every DVB-CPCM has a public/private key pair provided by an authority. Two devices exchange their signed certificates and, if not revoked, establish a Secure Authenticated Channel (SAC). The Content License is transferred through the established SAC. Since only DVB-CPCM compliant devices participate, content owners can be sure that the usage rights will be properly enforced.

Figure 9.7 illustrates the interaction of an Acquisition Point with a DRM and a Consumption Point. Before transferring the content, the Acquisition Point and the Consumption Point establish a Secure Authenticated Channel (SAC). The DRM hands over the content it previously protected. The handover of the content is mostly in the clear but in a secure environment (such as within a SoC). This only occurs if the DRM is authorized to do this. The DRM rights management layer forwards the expected usage rights. The CPCM rights enforcement layer defines a 128-bit random number that will serve as a key for the scrambling of the content. The rights translation layer passes the DVB-CPCM usage rights to the rights enforcement layer. This layer builds the Content License and signs it. It sends the Content License to the Consumption Point using the SAC. The Consumption Point's rights enforcement forwards the usage information to the rights management layer. It will possibly authorize the rights enforcement layer, which will provide the key to the content protection layer. This layer will descramble the content to be rendered and displayed. The establishment of the SAC enforces the trust. Only CPCM devices have the corresponding signed keys delivered by the authority.

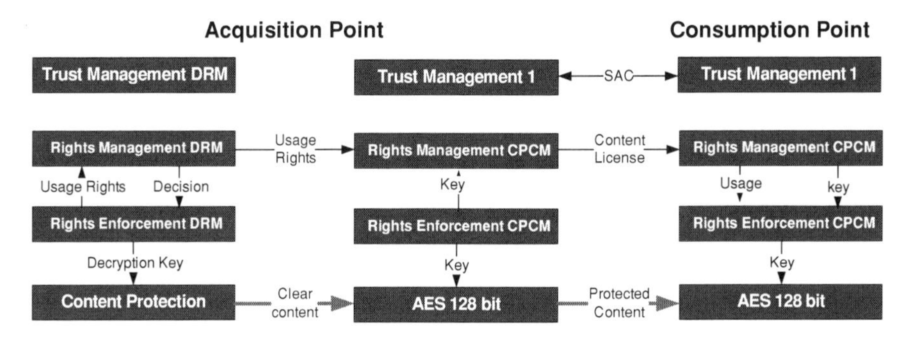

Fig. 9.7 Four-layer model for DVB-CPCM

9.6 The Analog Hole

Protection of digital content mainly relies on scrambling. This answer is suitable as long as the content remains digital. Unfortunately, the weakest link in the delivery chain is the analog link. Video has to be rendered on a display. Audio has to be channeled through loud speakers. Documents may be printed on paper, or displayed on screens. Pirates, like dishonest users, can play back the content, record the produced analog signals, digitize them, and redistribute the digital files. They can capture content using recorders with analog input, video grabber cards, camcorders, or cameras (Fig. 9.8). This method creates a hole in the digital content protection, called the analog hole. Some content distributors even describe how to exploit it to circumvent copy protection schemes [294]. Some companies use the analog hole to distribute content to other devices. For instance, doubleTwist [295] enables sharing of content on all of a customer's devices and with friends on social networks such as FaceBook. To circumvent FairPlay, doubleTwist uses the analog hole, i.e., it records content while it is being played by iTunes.

The only technological remedy available today against an analog hole is watermarking. The main idea is for content to carry a watermark that informs the recording device of possible copy restrictions. Prior to recording, the recording device looks for such watermark information and then decides if it should proceed with the recording. The initial thoughts were that the content should embed a watermark called CCI (Copy Control Information). CCI had four possible states:

- Copy free meant that there was no restriction on copying.
- Copy never meant that no copy was authorized.
- Copy once meant that a copy was allowed.
- Copy no more was the state of the instance produced by the duplication of "copy once" content. This state was problematic because it required modifying the CCI from "copy once" to "copy no more". In the case of a watermark payload, it required adding a new watermark. This approach required that a recording device also had a watermark embedder.

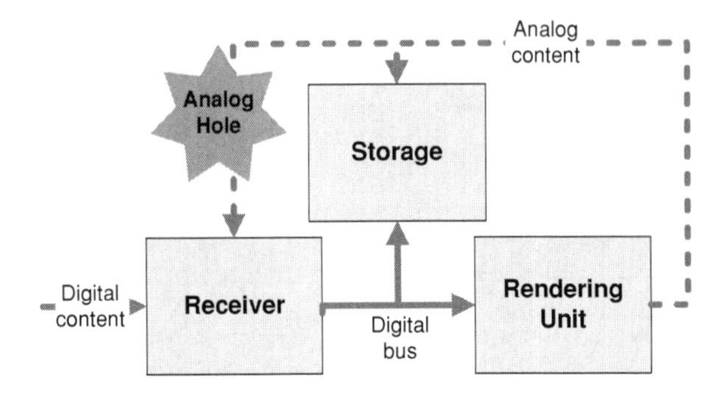

Fig. 9.8 The analog hole

A second approach proposed that the information carried by the watermark be just its presence. It is a zero-bit watermark [296]. A zero-bit watermark has the advantage of being more robust against attacks than a conventional watermark [297]. In the event of the detection of such a watermark, the recorder knows that it should find copy restriction information somewhere else, for instance, in metadata. In the absence of these data, the recorder terminates the recording.

Philips proposed a variant using the notion of authorized domain [298]. When content left an authorized domain, a domain watermark would be added. When it entered an authorized domain, the content would be checked for whether it carried a domain watermark. Were that the case, the piece of content would be rejected. The advantage would be that a piece of content would not need to be watermarked at the head end. This would simplify operations for free to air broadcasters. Nevertheless, this approach has a weakness. The system has to discriminate between content that should use the protection of an authorized domain and content that should not. For instance, exporting devices should not watermark personal video. Else, it would be impossible to view personal videos outside the domain of the owner.

In any case, the use of watermarks to transmit copy restriction information requires that all inputs of a recorder have a watermark detector. This means that this method is only efficient for compliant recorders, i.e., devices implementing the detector. Noncompliant recorders would not verify the watermark. Thus, watermark detection against analog holes requires legal enforcement, typically through robustness and compliance rules (Sect. 3.8, Compliance and Robustness Rules). These rules mandate the presence of the watermark detector with the associated behavior in every compliant device.

Obviously, this approach is prone to oracle attacks. In the event of an oracle attack, the attacker has access to an oracle that provides a Boolean answer if the challenge succeeds. In this configuration, the oracle is the watermark detector of the recording device and the challenge is to fool the detector. The attacker can alter content and then assess whether she succeeded in washing the watermark. If she can record, she won. Using several contents and different recording devices, she can design a class attack, i.e., one working on any device. The Secure Digital Media Initiative challenge (SDMI) [299], the first Break Our Watermarking

System contest (BOWS) [300], and the second BOWS contest (BOWS2) [301] highlighted the potential of this type of attack. Currently, no commercial product deploys one of these systems. Nevertheless, a watermark is embedded to cope with the analog hole. The embedded information is not CCI but some forensics data. The detection is a posteriori once an antipiracy team spots an illegal content. This strategy is not vulnerable to oracle attacks.

Chapter 10
Digital Cinema

In January 2000, the Society of Motion Pictures and Television Engineers (SMPTE) launched the DC28 working group dedicated to digital cinema. Subsequently in March 2002, seven major US studios, Disney, Fox, MGM, Paramount, Sony Pictures, Universal Studios, and Warner Bros., launched the DCI, whose goal was to accelerate the adoption of digital cinema by issuing specifications and recommendations, which would later be used in SMPTE. In 2005, they issued the first DCI specifications [95]. Currently, SMPTE "owns" and maintains the digital cinema-related specifications.

Digital cinema contents are transferred in containers called Digital Cinema Packages (DCP). DCPs are protected as illustrated by Fig. 10.1. The audiovisual essence and the optional subtitles are scrambled with AES-CBC using a 128-bit key. Each component may use a different random key. HMAC-SHA1 guarantees the integrity of each component by using the same key as the one used for scrambling. The scrambled components are packaged together in a DCP. The Material eXchange Format (MXF) defines the format and syntax of the DCP.

The descrambling keys are packaged in a license called the Key Delivery Message (KDM). Each KDM is dedicated to a given theater. To generate the KDM, the operator needs the list of the players that will play the DCP. The owner of the theater communicates this list to the authority generating the KDM using the Facility List Message. For each player, its manufacturer generates a unique RSA key pair, the corresponding X509 v3 certificate with the public key [79], and information such as the model and serial number. The Facility List Message holds all certificates of the theater's players. The authority generating the KDM must revalidate these certificates through CRLs or OCSPs [241]. Once the certificates are validated, the authority builds the Trusted Device List, which defines the devices within the auditorium that are authorized to access the DCP. The descrambling keys are encrypted with RSA using the public key of each authorized player. The KDM contains also usage rights that define conditions of playback, for instance, the authorized period of displaying and the number of authorized representations.

E. Diehl, *Securing Digital Video*, DOI: 10.1007/978-3-642-17345-5_10,

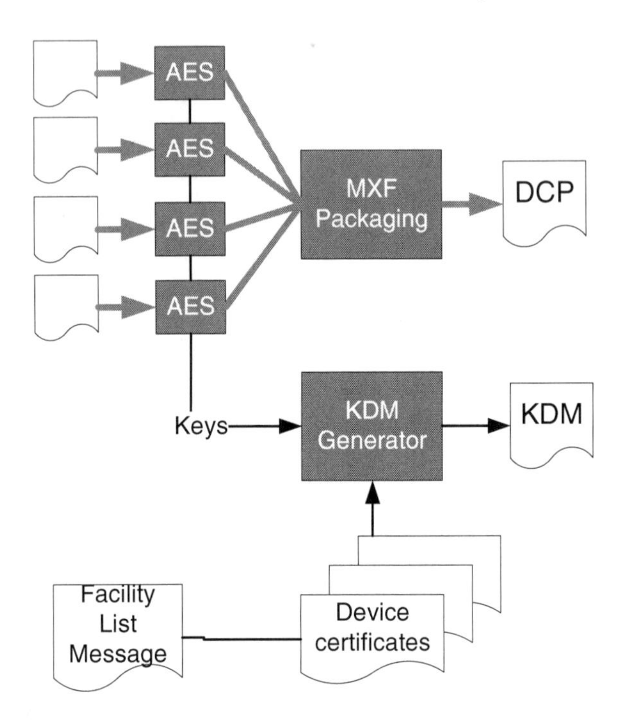

Fig. 10.1 Protecting DCP

Figure 10.2 provides a simplified view of the playback of a DCP. The video server delivers the DCP to the player. A media block handles the audiovisual essence. Within the media block, the DCP is depackaged. Each component is descrambled using the AES descrambling key delivered by the security manager.[1] The JPEG2000 video is decompressed.

Optionally, an invisible watermark can be embedded in the decompressed video before display (Sect. 3.3.3, Forensic Marking). The embedded payload uniquely identifies the player and the time and date of playback. The embedded message carries the time, date, and location of the projection. Additionally, each video server has a unique identifier. It is thus possible to trace the theater where the unauthorized camcorder capture occurred [302]. The specifications define minimal formatting requirements for the payload. The time stamp should have at least a resolution of 15 minutes. The location should be encoded on 19 bits. The entire payload should be repeated at least every five minutes. The SMPTE specifications provide a set of minimal requirements for the watermark, but do not define the technology itself. Currently, one product dominates this market: Civolution's NexGuard [303].

[1] To simplify the figure, we did not draw the subtitling components. They are of course managed by the media block. Similarly, we did not draw the video and audio treatments, except the JPEG2000 decompression.

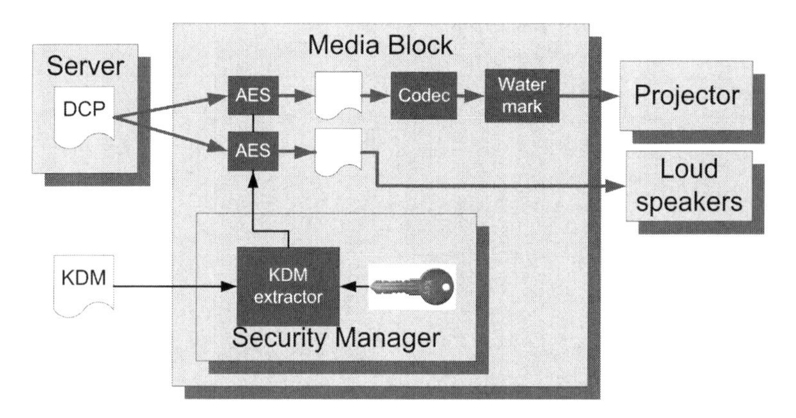

Fig. 10.2 Playing DCP back

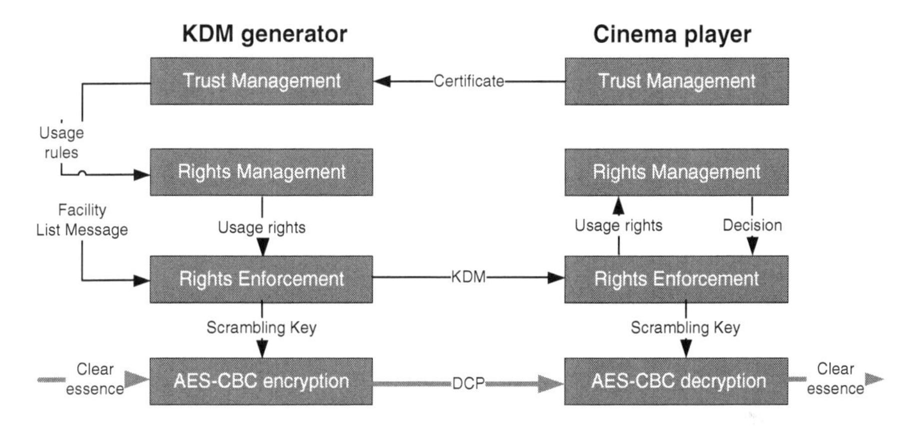

Fig. 10.3 Four-layer model of digital cinema

The media block has a dedicated secure component: the Security Manager. The Security Manager holds the private key of the player. It receives the KDM and decrypts it using this key. It validates usage rights such as authorized time windows. Furthermore, the Secure Manager securely logs all transactions. DCI requires the Secure Manager to be at least FIPS140-2 level 3-certified (Sect. 3.5.3, Dedicated Countermeasures).

Figure 10.3 presents the four-layer model of digital cinema. The trust management layer highlights the fact that the main element of trust is the authority generating the KDM. The system assumes that the certificates delivered by the Facility List Message have all been properly validated by this authority.

The transfer between the player and the projector is also protected by encryption using CineLink2, a link encryption designed by Texas Instruments.

All these elements are defined by a set of specifications. (Table 10.1)

Table 10.1 Digital cinema standards

Standard	
SMPTE 427	Defines the method for secure transfer between the player and the projector. It is a link encryption over a high-bandwidth connection
SMPTE 429-6	Defines the syntax and structure of DCP as well as the scrambling mode (AES) and the integrity code (HMAC-SHA1)
SMPTE 430-1	Defines the syntax and structure of KDM as well as the encryption
SMPTE 430-2	Defines the format of the digital certificates as well as the associated processes
SMPTE 430-9	Defines the KDM bundle which carries several KDMs for a given facility
SMPTE 20-2010	Defines the CineLink2 link encryption

The overall scheme protecting the digital cinema DCP is secure. The trust model relies on the authority generating the KDM, and on the manufacturers of equipment. The main point of weakness comes from insiders acting unlawfully [304]. Currently, there is no known hack of this system. The major threat remains camcording in the theater.

Chapter 11
The Next Frontier: Interoperability

11.1 Why Interoperability?

Many media channels bring content to consumers. It may be broadcast using terrestrial, cable or satellite channels. IPTV brings the same broadcast channels through the phone line. These channels, at least when they are not free to air, are often protected using CAS. Consumers may prefer selecting content rather than watching a broadcast program. They use pre-recorded content, as on DVD, or download content from VOD or subscription services. DRM systems protect these services. Unfortunately, there are many competing DRM systems and content protection systems. Most of them are not interoperable.

Consumers expect to have a seamless experience, as they did with analog technology. For instance, they could play their content on any of their devices. Today, this is not possible. Consumers still expect to play their content on any of their numerous appliances. They expect to be able to lend their content to friends and relatives. They expect to be able to resell or give away content they are not any more interested in. In the US, the first resale doctrine, codified in the US Copyright Act, Section 109, authorizes a user to transfer legally acquired content to somebody else without having to ask for permission. Currently, however, DRM systems may be a limiting factor to all these expectations and thus spoil the user experience. We now examine two scenarios that highlight the limitations of non-interoperating DRM systems.

Alice purchases mp3 songs from an online service. She downloads these songs to her favorite portable player. A few months later, Bob offers her a new, more sophisticated portable player from another brand. Happily, Alice transfers all her legitimately purchased mp3 songs to her player. Surprisingly, she cannot listen to her songs anymore. Why? The first portable player is equipped with DRM A, whereas the newer one uses DRM B. Although both portable players support the same audio format, mp3, they do not support the same license format and bulk encryption. In fact, when purchasing songs, Alice actually purchased DRM

E. Diehl, *Securing Digital Video*, DOI: 10.1007/978-3-642-17345-5_11,

A-protected content. Thus, if she wants to listen to her favorite songs on her new player, she has to purchase a copy protected by DRM B.

Alice has learned a lesson. She remains loyal to the brand of her player. She usually purchases her songs from the same online service run by merchant A. Recently, she has discovered a new artist. She looks for his songs on the site of merchant A. Unfortunately, merchant A does not distribute this artist. She searches on the Internet and finds a merchant B that distributes this artist. She purchases several songs from merchant B. When playing them back, she discovers that they do not play on her portable player. Why? Merchant A supports the DRM of her portable player, whereas merchant B does not support it.

These scenarios highlight some of the problems for consumers. However, the lack of interoperability affects all stakeholders, not only consumers.

With present solutions for DRM, consumers are constantly frustrated by a variety of issues:

- They have very restricted user experience allowed with legally purchased content, compared to the experience they had with analog content [305] or the experience they get from pirated content.[1] Most consumers are willing to pay for their content, though this is challenging when the playback of this content is restricted to one unique device or impossible to back up their content. A typical complaint is the impossibility of backing up DVD to preserve content, say, from children's sticky fingers. Eventually, the original DVD, full of scratches, will be unreadable.
- Old formats are not always supported by new DRMs. This means that legally purchased content may not be playable anymore because it has been made obsolete by the new DRM. Some frustrated customers may turn to illegal distribution channels that provide content playable on all devices [306].
- When multiple DRMs developed by the same company are not compatible. This is the case with Microsoft's Zune DRM and Microsoft's PlayForSure DRM [307]. A song purchased for a Zune device cannot be listened to on a Windows computer because their DRMs are not compatible!
- Legitimately purchased content is sometimes no longer playable on one of their devices. If the merchant no longer supports the DRM, for instance, after bankruptcy, then there is no way to get new licenses [308]. In some case, some licenses are linked to a unique appliance. If the configuration of this appliance evolves, or if the consumer transfers her content to another appliance (even one compatible with the DRM), there is no longer an available authority to update the licenses [309].

Device manufacturers choose between implementing just one DRM system or several DRM systems. The first case will produce a cost-efficient product, but the

[1] Pirated content does not (normally) have any copy protection or DRM. Hence, it can be played on any device, circumventing the interoperability issue. Furthermore, it is often free (or at least cheaper).

choice of the right DRM becomes crucial. The second case will produce a costly and complex product. Unfortunately, this device will not be future proof if a new type of DRM appears.

Distributors face the same dilemma as manufacturers do. Using a single DRM may limit their market share. Furthermore, due to network effects, distributors should select the DRM with the largest installed base rather than the DRM that best fits their requirements [310]. An alternative is to support several DRM systems. Nevertheless, this proportionally increases distributors' operational costs. Furthermore, if due to a technological choice, the distributor cannot access the catalog of some content owners, then the distributor will lose market share.

Consumers want to be able to access a large part of the world's media from any portal [311]. Content providers have access only to a fragmented market. If consumers do not have easy access to the content distributor's legal offering, they may look for alternate illegal sources. Furthermore, content distributors may need to supply content in different formats. Content providers have to find the right tradeoff between monetizing their content through versioning and the expectations of their customers. With non-interoperable DRMs, customers are unlikely to buy into the versioning strategies of studios.

DRM is a perfect illustration of an economic network effect [312]. The more a DRM solution is deployed, the higher is the potential market shares of the content it will protect. Of course, a serious commercial offering must support the scale of the deployment. Commercial success requires both a large, deployed base and a large offering. Some manufacturers, especially Apple, use the non-interoperability of the DRM as a strategic choice. FairPlay, Apple's DRM, is the single DRM allowed to protect content for the iPod/iPhone/iPad family. Furthermore, FairPlay-protected content is only available on Apple's website "iTunes". This strategy efficiently locks consumers to Apple's proprietary hardware [235, 313].

In order for DRM1 to interoperate with DRM2, there is the prerequisite that the technology provider of DRM1 and the technology provider of DRM2 agree to interoperate. Although obvious, this prerequisite is tough. Interoperability has to fulfill the following three goals [314]:

1. Essence protected by DRM1 has either to be compatible with formats and scrambling supported by DRM2 or to be securely converted to essence protected by DRM2. This corresponds to the content protection layer of the four-layer model. Currently, each DRM uses proprietary secure protocols and its own scrambling technology. A common misunderstanding is that if two DRMs use the same scrambling algorithm, for instance, 128-bit AES, then they are interoperable. Unfortunately, they may use it with different modes, a different initialization vector (IV), or different crypto periods, and thus be fully un-interoperable.
2. A license for DRM1 has to be translated into a license conforming to DRM2. All RELs used by DRM are not equally expressive. For instance, CCI of DTCP only uses two bits, whereas MPEG21 REL defines powerful expressivity [285]. The usage rights must be consistent among DRMs in a predictable and

understandable way for the customer. This requires interoperability at the rights management and rights enforcement layers.
3. DRM1 must be sure that the terminal hosting DRM2 is trusted before handing over protected content. This requires interoperability at the trust management layer. Today, most DRMs do not trust each other.

All these goals imply that content providers must now not only trust the protected format used to initially distribute their content, but also any other protection format into which that content may be later translated as well as any associated services, devices, and players the content may reach during its lifetime. This trust includes confidence that usage and distribution rights expressed in one system maintain their equivalent semantics when expressed in another [313].

If interoperability implies several DRMs, then the web of trust becomes rather complex. Trust is not necessarily a transitive operation. It is often a matter of negotiation among different stakeholders.

11.2 Different Types of Interoperability

There are many approaches for reaching interoperability [315]. Rob Koenen proposed an interesting taxonomy [316] with three categories: full-format interoperability, connected interoperability, and configuration-driven interoperability. In the first mode, it is expected that all devices share the same formats, C&R regimes, and usage rights. Connected interoperability assumes that consumers will have a permanent connection. Remote servers will handle the interoperability problems. Configuration-driven interoperability assumes that the tools needed to access the content will be downloaded when needed. We propose a richer taxonomy with five categories:

- *The vertical approach* solves the interoperability issue by defining one universal format and protection scheme used by every device. It is equivalent to Koenen's full-format interoperability.
- *The horizontal approach* solves the interoperability issue by defining some common interfaces and mechanisms while the definition of the remaining elements stays open. It only offers partial interoperability.
- *The plug-in approach* solves the interoperability issue by providing a framework that first locates and then downloads the tools needed to implement the actual DRM. Koenen calls it configuration driven interoperability.
- *The translation approach* solves the interoperability issue by defining a set of mechanisms that convert protected content from one DRM into another DRM. Koenen calls it connected interoperability.
- *The Interoperable Rights Locker* solves the interoperability issue by merging the horizontal approach with the translation approach. This is currently probably the most promising approach.

11.2.1 The Vertical Approach

The vertical approach is hegemonic. The assumption is that all devices support the same DRM. Therefore, interoperability is ensured at least among the set of devices supporting this DRM. All devices support the same four protection layers; thus, they can easily exchange content if authorized. This type of interoperability only works if all devices share exactly the same trust management layer, especially the same certification authority. All devices supporting the same DRM must mutually authenticate each other. This requires that all merchants share the same trust authority.

Two approaches exist:

- the proprietary approach;
- the standardization approach.

In the proprietary approach, one company exclusively designs the system and may sell its technology to third parties. The specifications remain secret, at least for the content protection and rights enforcement layers. The company has two options:

- The company may decide to manage the complete system. It does not allow any other company to implement its DRM. This is the case with Apple's FairPlay.
- The company may decide to sell its technology to an Original Equipment Manufacturer (OEM). In this case, only the API is public. Sometimes, the rights management layer is public. The designing company delivers the DRM Software Development Kit (SDK) to manufacturers that will package it in their own applications. Having a single technology supplier has the advantage that implementations will be inherently compliant with the standard. Furthermore, the risk is homogenous and controlled (or at least known). Microsoft Windows Media DRM is one example.

In the second approach, a consortium of companies designs the system. It makes the specifications available either publicly or through a non-disclosure agreement. Any company may design its own implementation of the DRM following the specifications. Thus, to ensure interoperability, it is mandatory that all implementations use the same trust management layer. Although two systems may implement the same specifications, if they do not share the same root of trust, they will not be interoperable. This means that they must share the same compliance and robustness regime. Ideally, an independent body should assess the compliance of every implementation. This requires the definition of a strict compliance procedure. The trust management layer requires credentials. Often, the successful certification of compliance is contingent on the delivery of these credentials. A common licensing authority delivers these certificates. OMA is one example [239] with CMLA.

In fact, vertical interoperability only offers limited interoperability. Interoperability only occurs among devices supporting the same vertical system. Different vertical systems do not interoperate. Furthermore, there are no common elements among different systems. Unfortunately, it is similar to the current situation. Vertical interoperability works if the market share of this DRM is extremely high or, ideally, quasi-hegemonic. In that case, a majority of devices will interoperate, providing a sense of interoperability. Consumers will blame the minority of devices that do not support the "hegemonic" system as being made by companies that do not play the game.

Thus, to be successful, vertical DRM must create a large ecosystem with millions of pieces of content, millions of users and, tens of millions of devices and applications [317]. This is, for instance, the case with OMA in the mobile phone market, Microsoft Windows Media DRM in the PC market, and FairPlay in the Apple iPod community.

11.2.2 The Horizontal Approach

In the horizontal approach, two interoperating DRM systems share at least one common layer. Nevertheless, the two interoperating DRM systems have at least one distinct layer. The DVB project defined the first historical interoperable DRM. DVB is an industry-led consortium of over 270 broadcasters, manufacturers, network operators, software developers, regulatory bodies, and other organizations in over 35 countries committed to designing global standards for the global delivery of digital television and data services. Services using DVB standards are available on every continent, with more than 120 million DVB receivers deployed [215].

In the early 1990s, Pay TV was still in its infancy. Many European broadcasters were competing for its market. Several CAS providers were competing for market share. Broadcasters were bearing both the cost of STB and the cost of the broadcast and head-end equipment. It was rapidly clear that price reduction could only happen with a mass market. Thus, it became necessary to standardize as many elements as possible. DVB is consensus-driven. Thus, a DVB-adopted solution is a finely crafted equilibrium that respects at its best the interests of all stakeholders. Thus, DVB standardized the format of the video, selecting MPEG2. DVB standardized the way the programs were signaled on the air (DVB SI). This was sufficient for FTA broadcasters but insufficient for Pay TV operators because no piece of security was yet standardized.

Pay TV operators wanted to have maximum compatibility, whereas CAS providers were afraid of too much standardization. The rationale of this fear was the risk of a class attack[2] defeating all DVB systems. Thus, the system had to provide means to simultaneously support different CAS technologies. After many years of collaborative work, DVB issued the DVB SimulCrypt [318] and Multi-Crypt standards. With SimulCrypt, several CAS technologies can protect each

---

[2] A class attack is an attack that is applicable to every instance of the hacked system.

stream. The stream simultaneously carries the ECMs and EMMs of each CAS provider. The STB embeds one CAS and selects corresponding information in the stream. With MultiCrypt, only one CAS protects each stream. The STB can embed several CAS technologies and select the right CAS. SimulCrypt puts the burden of cost on the head end side, whereas MultiCrypt puts the burden on the STB. SimulCrypt is widely deployed. In these standards, DVB fully defined the content protection layer as illustrated in Fig. 11.1.

DVB defined the scrambling algorithm used to protect the essence. It was the DVB-CSA. DVB-CSA uses a 64-bit key called Control Word.[3] The Control Word is valid for a given period, called crypto-period. DVB-CSA's crypto-period can vary from 10 to 120 s. DVB defines two types of licenses: ECMs, and EMMs. An ECM carries the Control Words together with the identifier of the program and usage rights for the rights enforcement layer. An EMM updates the consumer's rights in the rights management layer. DVB's definition of ECM and EMM is limited to the minimum requirements:

- the header to parse the stream and extract the right data packet;
- the size of the ECM and EMM; and
- the placeholders of the ECM and EMM within the signaling information.

DVB does not define the internal structure of the ECM and EMM, nor how they have to be protected. Each CAS provider manages them in its proprietary way. To allow this proprietary handling of ECMs and EMMs, it was decided to export their execution to smart cards.

This DVB design decision was an extremely wise one from a security point of view. Designing a robust scrambling algorithm is a difficult but not impossible task. A group of European cryptographers performed this delicate task. The specification of DVB-CSA is not public. It is only available through a non-disclosure agreement from ETSI. Until 2002, the specifications of this algorithm did not leak. In the fall of 2002, thanks to reverse-engineering, a pirate software-based decoder `FreeDec` provided access to the DVB-CSA specifications. Currently, there is no serious cryptographic attack on DVB-CSA [218, 319]. Nevertheless, the initial size of the key (56-bit) became too small when compared to current recommendations [320]. Fortunately, the crypto-period is less than 120 s. This means that it is not yet feasible to brute-force DVB-CSA protected content. Nevertheless, DVB defined new versions of DVB-CSA using longer keys.

Recently, leading CAS providers developed a new framework: OpenCA [321]. This framework uses the same elements as DVB but targets a larger scope. For instance, within the mediaFLO system, it can be used to carry any content over 3G networks. The system allows sharing the same Control Words for the same content protected by different CASs.

Designing a conditional access system is more complex than just designing a scrambling algorithm and thus the former is more prone to attacks. Many

[3] DVB CSA3, the latest version, uses 128-bit keys.

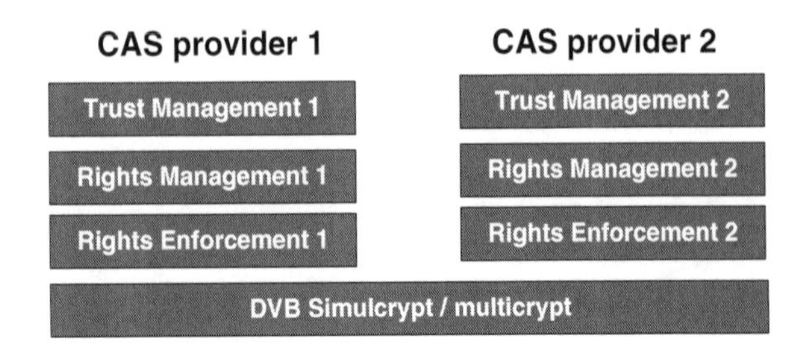

Fig. 11.1 DVB interoperability scheme

successful attacks occurred against deployed CASs, mainly on the implementations. Nevertheless, because a smart card hosts the conditional access, it is relatively easy to cope with the attacks by replacing the smart cards.[4] Each new generation of cards implements a newer version of the CAS system more robust than the previous one. Furthermore, the newer smart card benefits from the latest progress in tamper-resistance.

DVB is a success story. Many Pay TV operators have used it for many years all over the world. The DVB approach proves that using one common format and one common scrambling algorithm for the essence does not reduce the overall security of the system. It reduces the investment cost by lowering the price of the hardware and reduces the operational costs.

The Devil's in the Details

One of the most important security advantages of smart card-based architecture is the renewability of the overall security. If the designer exports all the security features to the smart card, then recovering from a new vulnerability is easy. Changing the old smart card for a renewed and patched smart card should address the vulnerability. This was the paradigm used by the Pay TV systems of the 1990s.

Nevertheless, the designers of this architecture neglected a weak point: the path of the control word from the card to the STB. John McCormac disclosed an attack: the jugular hack [199], also known as McCormac's hack. In the early 1990s, the communication between the host and the smart card was not secured; hence, the control word was returned in the clear. McCormac proposed eavesdropping on the communication between the card and the host to capture the control words and broadcast them. Several customers could share a single subscription. When disclosed, this attack was purely theoretical. It required setting up a radio broadcast system, which could have been easily traced. The Internet was not yet widely deployed.

When designing the first Pay TV systems, the choice of returning clear control words was acceptable. With the advent of hard disk and the Internet,

[4] The replacement of the smart card can be triggered by economic factors. As long as the estimated loss due to piracy does not exceed the cost of replacing all the smart cards in the field, it is useless to replace them.

the context changed. Instead broadcasting the control words over the air in real-time, Alice could just sniff all the control words exchanged with the smart card for one event, store them in a small file, and share this file over the Internet. McCormac's hack changed from a theoretical attack to potential, practical attack. Sharing a control words became a commercial, illegal service [322]. Bob stores the scrambled essence on his hard disk and retrieves, from the Internet, Alice's file containing all the corresponding control words. This operation minimizes the required bandwidth for piracy. It costs Alice less to send some control words rather than the full movie in the clear.
Current systems have implemented two countermeasures:

- The smart card and its STB establish a secure authenticated channel before exchanging control words. This channel offers two advantages:
  - The smart card passes control words only to an authenticated STB, and not to a rogue one. Revocation mechanisms allow discarding pirate set top boxes and computer emulators.
  - A session key encrypts the control word during the transfer between the smart card and the host. Sniffing the communication is useless.
- Systems implementing PVRs or home networks are very convenient for McCormac's hack. Thus, some systems rescramble or super-scramble [323] the essence before storing it on a hard disk or sending it to the home network. Rescrambling means that the system first descrambles the incoming essence and then immediately scrambles it using a different control word than the one used by the native scrambling. The rescrambling algorithm may be different from the native one. Super-scrambling means that the system immediately scrambles the already scrambled content using a different control word than the one used by the native scrambling. The super-scrambling algorithm may be different from the native one. In both cases, if Alice retrieves the control words protecting her stored "Shrek 28", it is useless for her to send them to Bob. Bob's "Shrek 28" is protected by other control words.

Lesson: List and document all security assumptions used when designing a system. All evolutions of the system require verifying that the initial assumptions are still valid [192]. In the 1990s, sending back control words in the clear was reasonably secure. The benefits for the hacker were small. Ten years later, this hypothesis went wrong, requiring the redesign of many Pay TV systems.

Another possible approach to horizontal interoperability is through a common REL, as illustrated in Fig. 11.2. The internal format of the license is identical for all DRM providers. The protection of the essence remains different. Some RELs are extremely complex and expressive. They may have to express complex usage

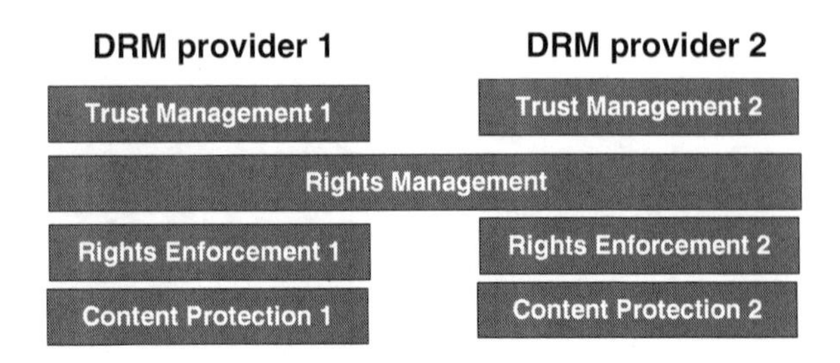

Fig. 11.2 REL interoperability scheme

rights. The Rights Enforcement Layer may not support the full set of these expressed usage rights. In that case, the Rights Management Layer has to translate from the non-supported usage rights into supported usage rights. There are two possible strategies: either to refuse the unsupported usage rights, or to define a closer usage right (Sect. 11.4.2, Usage Rights Translation). A consistent strategy is recommended, especially when more than two DRMs interoperate. The consumer should expect consistent behavior from all interoperating DRMs. She should have the same usage rights for a given piece of content regardless of the DRM.

Although often praised by scholars, this approach is of limited interest. It is of little interest to the B2C market. First, every device has to implement proprietary rights enforcement and content protection layers. The rights enforcement layer is the most complex one, and thus the most prone to vulnerabilities. The simpler the design is, the more secure the system will be. Furthermore, in B2C, the usage rights must suit attractive commercial offers. Here also, the trend is towards simplification. Nevertheless, this approach may be worthwhile taking in B2B. Merchants, which propose content protected by different DRM systems, would benefit from using a common REL. The definition of the commercial offer and its translation into usage rights become then common, simple tasks. Furthermore, the trading of commercial rights in the B2B market is far more complex than in the B2C market. Having to master a single REL would be valuable to practitioners.

11.2.3 The Plug-In Approach

The principle of interoperability through plug-ins is extremely simple. Protected content carries information about the tools needed to access it. If the DRM client, called terminal in this section, does not yet have these needed tools, then it knows from where to download them.

Thus, interoperability through plug-ins requires the definition of at least the following elements:

- a method to declare the tools needed to access the content and to define how to use them;
- a method to declare the location(s) where these tools may be found; and
- a method to download one or several tools and execute them.

A typical simplified scenario is as follows:

1. The terminal retrieves the protected content and its associated metadata, which carries the information described previously.
2. The terminal extracts the list of needed tools from the metadata.
3. The terminal checks if it already has the required tools using the description of the tools.
4. If this is not the case, the terminal identifies the location(s) where to find the missing tools. For instance, the metadata may provide one or several URL(s) where the tool is available.
5. The terminal downloads the missing tool(s) from a signaled location.
6. To successfully access the content, the terminal executes the downloaded tools in the recommended way (Fig. 11.3.)

In theory, interoperability through plug-ins is full-fledged interoperability. A terminal would just need a framework for the signaling, the downloading, and the execution of the tools. This vision is based on a set of assumptions:

- the terminal has access to a broadband connection; some tools may be complex and thus have large executable program;
- the tools are purely software based and do not require specific hardware;
- the tools are available for all deployed platforms; and
- one single certification authority manages a common trust layer.

The first assumption is probably valid, especially in our connected world. The second assumption may introduce some strong constraints. It may exclude tamper-resistant hardware, such as smart cards or SoCs. A tool that would require dedicated hardware may not be supported by all platforms, unless dedicated hardware is mandatory in the standard. There are three ways to fulfill the third assumption: fully specify the terminal, specify a virtual machine (such as Java), or port the tools to every targeted terminal. The first approach is sometimes called closed garden or walled garden.

As with all DRM solutions, one difficult challenge is compliance with the robustness and compliance rules. These rules are necessary for achieving complete protection (Sect. 3.8, Compliance and Robustness Rules). In a closed garden approach, it is relatively easy to enforce compliance. The specification of the terminal defines robustness rules that the terminal must support. For the two other approaches, it becomes more difficult. If the targeted terminal is a generic computer such as a PC, then it becomes extremely difficult.

Ideally, a tool should declare a set of requirements, a minimal performance threshold, and compliance rules that the downloading platform should comply with. Then, it should be checked whether the requirements are fulfilled. There are mainly three possible methods for this:

- The target terminal checks the requirements and validates its compatibility. Obviously, the trust model would require that the target terminal honestly

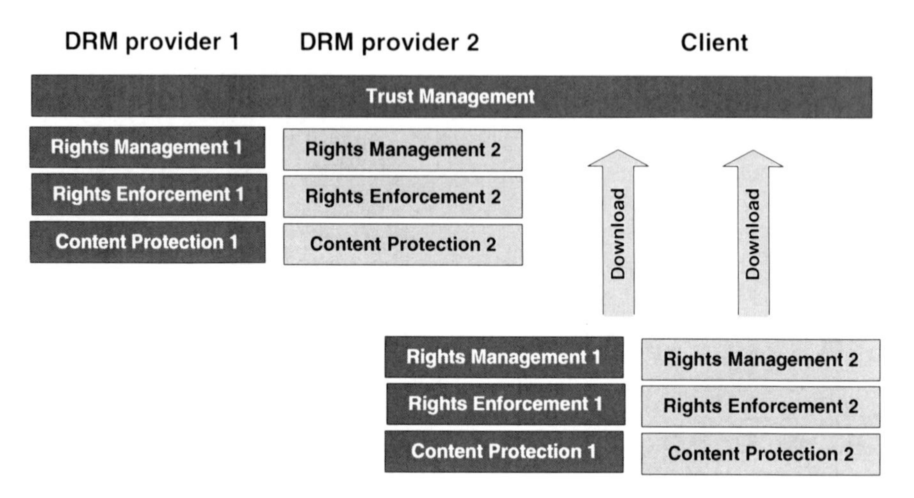

Fig. 11.3 Plug-in interoperability

perform this verification. This would not be the case if an attacker controls the terminal. The attacker may bypass this first step.

- The target terminal declares its capabilities. The tool evaluates whether the terminal is suitable or not. Here also, the trust model assumes that the target terminal makes an honest declaration. Using device certificates is a typical method that partly overcomes this limitation. A device certificate is a piece of digital information delivered by a trusted authority. The information may include description of capabilities, a compliance flag, and requirements. The trusted authority validates the veracity of the provided information and signs the device certificate. The tool evaluates the information and the validity of the signature.
- The tool performs tests to validate the target terminal. Here, the trust model is more robust because it does not make any assumption about the target terminal. In fact, an attacker may control the terminal, but she will have to successfully answer the tool's challenges. The trust model assumes that the tool will be able to evaluate the capabilities of the terminal. This may be possible for evaluating capabilities such as processing power, and available memory. Nevertheless, it is impossible to evaluate compliance and robustness constraints such as the analog output copy protection.

With interoperability through plug-in, the problem of trust is bilateral. As seen previously, the plug-in must be sure that the host is compliant. Furthermore, the host must also be sure that the plug-in is compliant and behaves safely. Without this precaution, the plug-in may become a vector of dissemination for malware. This trust establishment is the field tackled by mobile code security. One of the most exciting researches in this domain is Proof Carrying Code. George Necula and Peter Lee were its initial pioneers [324]. Proof Carrying Code is a technique by which a host can verify that a piece of code delivered by a non-trusted source fulfills a set of security requirements called safety policy. The producer of the plug-in has to generate a formal safety proof that its plug-in respects the safety policy of the host. The safety proof is delivered together with the plug-in. Prior to executing the plug-in, the host will verify the safety proof using a proof validator. Only once the proof validated can the host execute the downloaded plug-in. In complex environments such as the one of interoperable DRMs, trust is paramount. Being transparently able to prove the security of a plug-in would be an asset to consumers.

The most famous examples of the plug-in approach are MPEG4 Intellectual Property Management and Protection (IPMP) [325] and Digital Media Project (DMP) (Sect. 11.3.5). The market has accepted neither of these solutions. Currently, there are no commercial deployments.

11.2.4 The Translation Approach

The translation approach, sometimes called the intermediation approach, is conceptually simple but hard in practice. Device A (respectively B) only supports DRM A (respectively DRM B). Device B is requested to play back content protected by DRM A. Device B asks a trusted agent to translate the piece of content protected by DRM A so that DRM B protects it. First, the trusted agent verifies whether the action is authorized, i.e., whether device B can be granted access to this piece of content. If so, in its secure environment, the trusted agent translates the license of DRM A into the equivalent license of DRM B. If necessary, it descrambles the content and rescrambles it with the bulk encryption of DRM B. The trusted agent returns the same content, but protected by DRM B (Fig. 11.4).

The translation approach places the burden of translation on the trusted agent. From a security point of view, this decision may be wise. The trusted agent is usually a remote application executing on a server. We may assume that the execution environment of the trusted agent is well controlled, and thus easier to secure. The translation is of course a point of weakness, especially if conversion from one bulk encryption to another occurs. It should be simpler to protect it in a professional environment than in a consumer environment.

The translation approach has three drawbacks:

- translation of usage rights;
- combinatorial complexity; and
- trust management.

DRM A and DRM B may not necessarily use the same REL. Moreover, even if they share the same REL, their respective rights enforcement layers may not support the same usage rights. The translation has to map some usage rights to the closest equivalent usage rights. This translation is not trivial. Section 11.4.2 describes the problem.

The trusted agent may be of moderate complexity when managing interoperability between two DRMs. Complexity increases when the trusted agent manages

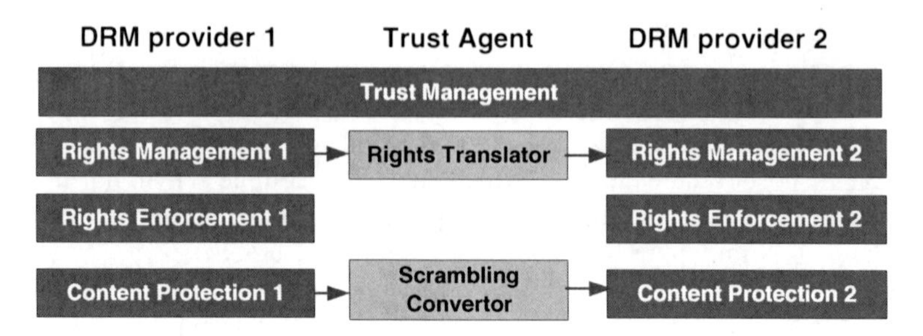

Fig. 11.4 Interoperability through translation

more than two DRMs. For instance, if the trusted agent handles n distinct DRMs, then it must implement up to $n\ (n-1)$ translation modules. Each module translates from one DRM into another DRM. The translations may be less than this number because interoperability is not necessarily a bijection. If DRM A "forwards" content to DRM B, this does not systematically mean that DRM B conversely "forwards" content to DRM A. For instance, compliance and robustness rules may forbid it.

The problem of trust management applies to different parts of the system. Obviously, all the content providers that expect their content to interoperate must trust the trusted agent. This global trust arrangement is not a technical issue but a legal and commercial one. Once this global trust is established, there must be the establishment of local trust between the two DRMs. In fact, the trusted agent must be able to open any license from DRM A, and to generate a new license for DRM B. A simplified trust model assumes that each content provider equally trusts all DRMs, which is not always the case. One content provider may fully trust DRM A, while not totally trusting DRM B. For instance, the content provider could decide to offer two types of content: premium content and regular content. Premium content would be available earlier than regular content. The content provider may agree to deliver all content when the content is protected by DRM A, but only regular content when the content is protected by DRM B. In this case, the trusted agent should reject any transfer of premium content from DRM A to DRM B. Another content provider may equally trust the two DRMs. In that case, the trusted agent should apply no restrictions.

The Devil's in the Details

This problem of limited trust in a DRM may seem artificial. Nevertheless, it is serious and real. In 2006, DVB defined a new version of the DVB Common Interface (DVB-CI). DVB-CI is an interface used to plug in a removable module using the PCMCIA form factor (Sect. 6.5, DVB-CI/CI+). The DVB-CI host transfers digital stream to the DVB-CI card. The card processes the stream and returns the modified stream to the decoder. Its initial purpose was to support Simulcrypt. Thus, the same decoder could handle different CAS systems. In the initial version, the stream was returned

in the clear. A few years later, this clear returned stream was identified as a place to easily perform digital copies. One of the purposes of this second version was to protect this stream.
Thus, DVB-CIv2 should return a scrambled stream rather than a clear stream. The standard has to define the scrambling algorithm. The members of DVB proposed two candidates: DES and AES. AES was the obvious candidate. For several years, DES was considered aging [326]. Nevertheless, many companies wanted to keep DES in order to be earlier in the market. Chipsets implementing AES would require at least 18 months to be ready, whereas chipsets implementing DES were available. A hot debate on the security of the DES interface occurred. The main argument of the proponents of DES was that using a short crypto-period of 10 s should be safe for the next 20 years. The result was a secession of some companies that created CI plus.

Coral is the most known example of the translation approach; see Sect. 11.3.1.

11.2.5 Interoperable Rights Locker

The Interoperable Rights Locker (IRL) is the latest trend in the interoperability field. Thus, we may assume that it inherits from the best breeds of each of the previous initiatives. The Interoperable Rights Locker is mainly anchored to three pillars [327]:

- the digital rights locker, an online repository of the usage rights of each piece of content acquired by the customer;
- a single Rights Expression Language that describes these usage rights; and
- the single interoperable format, a common file format for the media essence describing the container, the compression formats, and the scrambling algorithm(s).

Figure 11.5 illustrates the principle. A common content protection layer protects each piece of content. In other words, every client implements the same descrambler (and possibly forensic tools). The digital rights locker holds for each customer the usage rights for each of her acquired pieces of content. Furthermore, it holds for each piece of content the keys that were used to scramble it, or at least information enabling the calculation of these keys. Each client uses its own proprietary DRM license system.

The following is a simplified explanation of the IRL principle. To use an IRL, a content owner has first to be enrolled in the digital rights locker. Once enrolled, the content owner can register any of its content with this digital rights locker. When a piece of content is registered, a secret key, hereafter called Control Word (CW), is randomly generated. CW is securely sent to the packager, which scrambles the

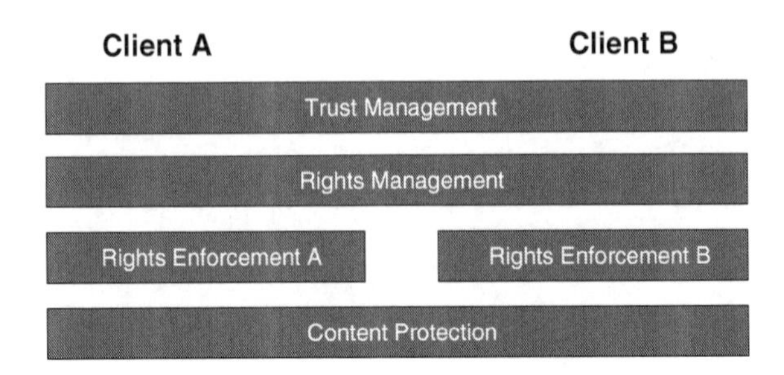

Fig. 11.5 Four-layer model for interoperable rights locker

content with the CW. CW is securely stored in the digital rights locker with other characteristics of the content.

To take advantage of the digital rights locker, a merchant needs to be enrolled in it. Once enrolled, a merchant may ask to register a customer (if she does not yet exist in the digital rights locker). Once a customer registered, the merchant can create or update the consumer's rights. A right defines the targeted content and its associated usage rights. The usage rights associate a piece of content with its access model, such as subscription, e-sell through, video on demand, rental, play once, and so on. Usage rights may also define the set of devices authorized to access the piece of content (Fig. 11.6).

Let us assume Alice purchased the rights to watch the movie "M" on all her devices. This is referred as electronic sell-through (EST). She owns a mobile device that supports DRM A and a computer using DRM B. The merchant registers with the digital rights locker the rights that she purchased the movie "M" as an EST.

Alice then receives a protected media file containing the movie "M". When she plays the movie back on her mobile device, the DRM A client requests a license using a URL obtained from the media file or a default URL. The URL points to the digital rights locker with information about Alice and her rights. Given that the request comes from Alice's mobile device, the digital rights locker identifies the type of supported DRM. The digital rights locker requests the license server A to generate a license for Alice's mobile device with the usage rights and the corresponding CW of the movie "M". License server A packages this information into its proprietary license format and sends it to the mobile device. DRM A may use a REL different from the REL used by the digital rights locker. The DRM client A opens the license, validates the user and the usage rights, and, if entitled, provides the CW to the descrambler. Now, Alice can enjoy her movie "M" on her mobile device.

When Alice wants to play the same movie "M" back on her PC, she does not necessarily need to download it again. She may use the media file she used with her mobile device. The DRM B client requests a license from the digital rights locker. As previously, the digital rights locker requests the license server B to generate a license for Alice's computer with the usage rights and the CW of the movie "M". License server B packages this information into its proprietary license format and sends it to Alice's computer. The DRM B client opens the license,

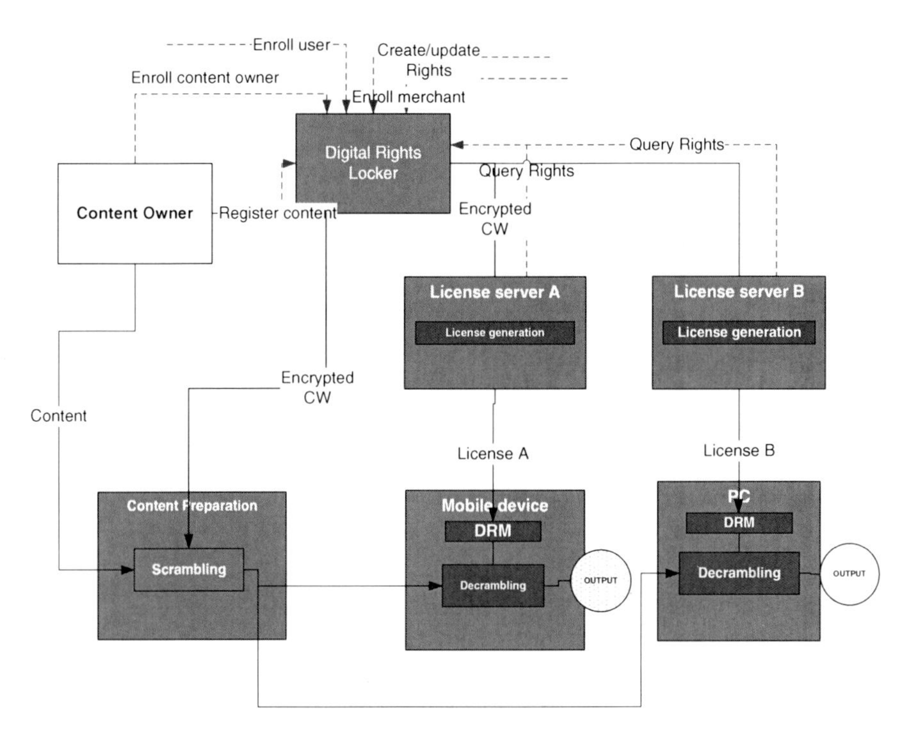

Fig. 11.6 Simplified schematic of the interoperable rights locker principle

checks the rights, and, if entitled, provides the CW to the descrambler. Now, Alice can watch the movie "M" on her computer.

Now that the main flow is established, let us describe a compelling use case that demonstrates the versatility of interoperable rights lockers. Let us assume that a movie is offered in an early release window at a premium, and after a few months, it is made available in a traditional release window such as Pay TV (see Fig. 2.2). The content protection requirements for the two offerings will most likely differ in certain aspects. DRM systems used by viewing devices having access to premium content should be the most robust ones. For illustration purposes, let us suppose that DRM A meets the security requirements of the early release window, whereas DRM B only meets the requirements of the latter window.[5] The movie is packaged using the single interoperable format. During the early release period, the digital rights locker only allows the license server of DRM A to issue licenses. Once the movie becomes available in the second release window, the digital rights locker allows the license servers of DRM A or DRM B to issue licenses. The transition from premium content to regular content does not require repackaging, thanks to the single interoperable format as well as the control of the digital rights locker.

[5] It is understood that DRM A also meets the requirements of the latter release window.

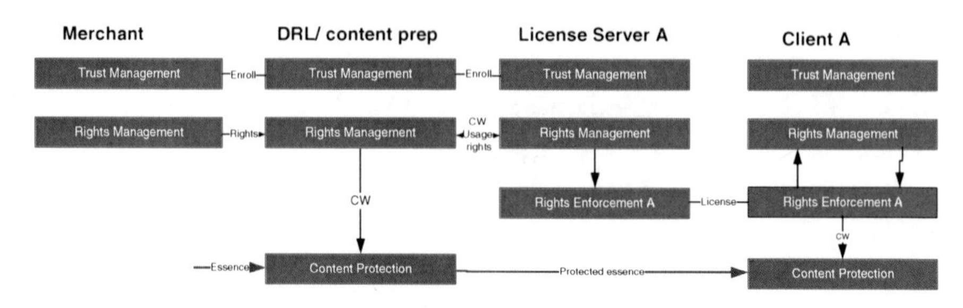

Fig. 11.7 Interaction among IRL actors

Figure 11.7 provides a different view of the interaction among the different actors adding the merchant. The merchant has to enrol in the digital rights locker to participate. Only enrolled merchants can communicate with the digital rights locker. Similarly, only enrolled license servers may query rights and obtain control words. The trust management layer ensures granular management of all authorized parties. Furthermore, it facilitates secure exchange of information among the actors.

The IRL architecture offers many advantages, among them the following:

- Trusted storage and management of customers' rights: Customers can thus access the same content on a large variety of devices, considerably enhancing their user experience.
- Simplified preparation of assets across the distribution chain: This reduces the operational cost for content distributors.
- Flexibility in usage rights and business models, allowing a potentially richer commercial offering.
- Ease of implementation for device manufacturers: They can select the DRM of their choice without decreasing the attractiveness of their devices.
- No dependency on DRM providers: Allowing access to acquired content to survive would make any provider disappear.
- No dependency on merchants: Access to acquired content is guaranteed to survive even if one merchant ends its activity.
- Support for untethered (off-line) consumption
- Reduced network congestion: Using a single file format eliminates the need for multiple downloads of the same piece of content for different devices.

Thus, the IRL is most probably the most flexible and powerful architecture for reaching DRM interoperability while being transparent for consumers. Nevertheless, to succeed this approach has to gain the support of many device manufacturers, of the majority of content owners, and of distributors and has to be ubiquitous [328].

There are currently two competing solutions; KeyChest (Sect. 11.3.4) and UltraViolet (Sect. 11.3.3).

11.3 Current Initiatives

11.3.1 Coral

Coral is a consortium of several tens of members. The founding members are Intertrust, Philips, Matsushita, Samsung, Sony, and Twentieth Century Fox Film Corporation [329]. Since October 2004, many new members have joined the consortium, such as NBC Universal, Sony BMG, Universal Music Group, IFPI (the organization representing the recording industry worldwide), DMDsecure, NDS, Pioneer, Seagate, and STMicroelectronics. The goal of Coral is to deliver an open standard for interoperability among DRM technologies for consumer devices and services. Coral issued its first specifications in November 2005 and the final set in October 2007.

Coral is currently the most elaborate example of interoperability through translation (Sect. 11.2.4). Coral interoperability is built on top of two main elements:

- NEMO secure messaging framework (Sect. 7.6, Marlin); and
- a pivot description of usage rights, so-called Rights Token.

In order to solve the issue of translation of usage rights, Coral uses a pivot format using the Octopus framework. Every piece of Coral-protected content has a 'license' described in a common language. The license, named *Rights Token,* contains the following mandatory information: *Principal P*, which identifies the node, *Resource C*, which defines the target content, and *Usage model U,* which defines the usage rights. The Rights Token expresses that principal *P* can access resource *C* under the conditions defined by *U*. The Rights Tokens supports a list of principals and list of resources.

Coral splits the handling of content into two phases [330]:

- The acquisition phase: Alice purchases content *C* using a Rights Token. The content provider generates a Rights Token that describes the usage rights *U* associated with the content *C* for a device or a set of devices *P* that Alice registered.
- The instantiation phase: When Alice wants to watch Coral protected content *C*, she asks for an instantiation of the associated Rights Token, i.e., a license compatible with the DRM of the targeted player. The same content may be accessible through devices using different DRMs, each using its own license format. For each DRM, the Rights Token is translated into corresponding native licenses.

A rights locker[6] stores and updates these Rights Tokens. A rights mediator performs the actual translation of Rights Tokens into native licenses for a given

[6] CORAL's rights locker is less sophisticated than the digital rights locker of the Interoperable Rights Locker concept (see Sect. 11.2.5).

DRM. It may be a remote online service. The rights mediation requires the following steps:

- The principal resolution checks that the target node belongs to the principal *P* defined by the Rights Token.
- The DRM verification: the rights mediator checks that the target node has a DRM that supports the usage model defined by *U*. The resolution of this problem is a complex task detailed in Sect. 11.4.2.
- The content resolution looks for an available instance of content *C* in a format supported by target device *P*. If it is not found, entities known as Resource Exporter, Resource Transformer, and Resource Importer transform the protected essence into a new protected format supported by target device *P*.
- The Rights Token resolution translates *P*, *C*, and *U* into namespaces understood by the DRM of the target device.
- The license creation transfers this information to a service called Rights Instantiator that generates the native license for the targeted DRM. This service is independent of the Coral system and is dependent on the native DRM. Thus, Coral does not assume that the target DRM is collaborating with and/or is able to provide information about the final native license. Coral only assumes that content *R* is available in a format supported by the targeted DRM.

Rights mediation can also occur locally in a home network, provided the right Coral roles are available [331].

Coral introduces a key concept: the ecosystem. The Coral core architecture provides a toolkit of principals with baseline behaviors. The Coral core architecture does not constrain, nor define, the usage models. Within an ecosystem, the specifications of the principals are refined to create a fully interoperable system and implement controlled usage models. The ecosystem defines the rules driving the instantiation of Rights Tokens.

Coral has defined an initial ecosystem (Ecosystem-A). The main concept behind Ecosystem-A is the interoperable domain. When acquiring Ecosystem-A content, customers should be able to play it back and share it freely on all devices belonging to their domain.

For that purpose, Coral defines new principals, as illustrated in Fig. 11.8. The user has one or several accounts. An account is associated with a Domain. A Domain encompasses a limited number of clients. A client is a device owned by the user. A client only belongs to a single Domain. The Domain Manager enforces the policy defined by Ecosystem-A. The user can add or remove a client from her domain by registering the action with her Domain Manager Service. When acquiring domain-bound content, the Rights Tokens are bound to the domain account. They are in the format (*D*, *C*, *U*) where *D* is the domain, *C* is the piece of content, and *U* is the set of the usage rights. The online rights locker manages the domain account. The combination of the domain manager and domain account solves all the namespace issues to generate the license for the devices belonging to the domain.

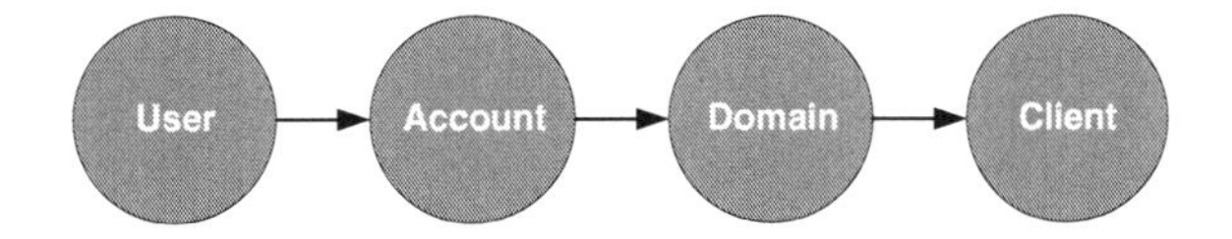

Fig. 11.8 Coral principals for domain management

Although the Coral consortium had many influential supporters, there are no commercial deployments. Nevertheless, Coral designed the foundations of Ultra-Violet (Sect. 11.3.3).

11.3.2 Portable Interoperable File Format

Microsoft announced Portable Interoperable File Format (PIFF), together with its Internet Information Services (IIS) smooth streaming transport protocol, at the International Broadcast Convention on 7 September 2009. PIFF's specifications [332] are open to all, on a royalty-free basis [333], under Microsoft's Community Promise.

PIFF is a container format that supports both file formatting and streaming. It is based on the ISO 14496 base media file format [334]. To support multiple DRMs, PIFF defines three new dedicated boxes uuids[7]:

- The Protection System-Specific Header box contains all information needed to play back the essence protected by one given DRM. There are as many Protection System-Specific Header boxes as supported DRMs. Each box contains a system ID identifying the corresponding DRM, the size of the data, and the proprietary objects needed to recover the decryption key. The proprietary objects are opaque to the system, which forwards them to the corresponding DRM client. The client DRM uses the proprietary objects in its own way.
- The track encryption box contains default values for the algorithm ID (clear, AES 128-bit CTR, or AES 128-bit CBC), the size of the Initialization Vector (IV), which may be eight or 16 byte long, and the Key Identifier (KID) for the entire track. The KID is a key identifier that uniquely identifies the needed decryption key.
- The optional sample encryption box holds the actual values of the IV and the KID and overwrites the default values defined by the track encryption box. It may apply to an entire track or to segments.

When playing PIFF content back, a PIFF player performs the following operations:

- It reads the Sample Description table to check if the content is encrypted.

[7] Universally Unique Identifier (uuid) is a 128-bit identifier standard defined by the Open Software Foundation. It enables distributed systems to consistently, and uniquely identify information.

- It consults the Protection Scheme Information Box to determine which DRM system handles the decryption key.
- It checks if the device supports any of the DRM systems listed in the Protection Scheme Information Box. If this is the case, it looks for the corresponding Protection System Specific Header box and passes the corresponding opaque proprietary objects to this DRM. The DRM uses corresponding information such as the license in its own proprietary way.
- The PIFF player uses the KID of the sample encryption box, if present, or the tracks encryption box to query the corresponding decryption key from the DRM. If access is granted, the DRM returns the decryption key. Knowing the decryption key, the decryption mode (CBC or CTR), and the IV, the PIFF player descrambles the piece of content.

With the PIFF approach, it is possible to develop a generic PIFF player that is DRM-agnostic. Indeed, PIFF defines the content protection layer of the four-layer model.

At the time of writing, PIFF is only supported by Microsoft PlayReady. It has been endorsed by Netflix [335]. It is expected that PIFF may become part of the UltraViolet specifications.

11.3.3 Digital Entertainment Content Ecosystem and UltraViolet

UltraViolet is a US initiative backed up by a large consortium of nearly 60 companies including Adobe, Alcatel-Lucent, Cisco, Comcast, DIVX, Fox Entertainment, Intel, IRDETO, Microsoft, NagraVision, NBC Universal, NetFlix, Panasonic, Paramount, Philips, Sony, Technicolor, Toshiba, and Warner Bros. It is interesting to note the absence of key players: Apple, Disney, and Walmart. The initiative started in 2007 with a project called OpenMarket [336]. In 2008, the initiative went public under the new name Digital Entertainment Content Ecosystem (DECE). Since 2010, it has promoted its system under the brand name UltraViolet.

UltraViolet (UV) attempts to reconcile all the stakeholders:

- Consumers can purchase content from any participating retailer and watch content seamlessly on any device that implemented one of the approved DRM systems. Consumers should be able to watch content in one place regardless of where they bought it.
- Participating retailers can sell content on behalf of content owners without restricting its market to a given brand of devices or a given DRM.
- Participating device manufacturers are ensured that their compliant devices offer access to a large collection of content proposed by many retailers.

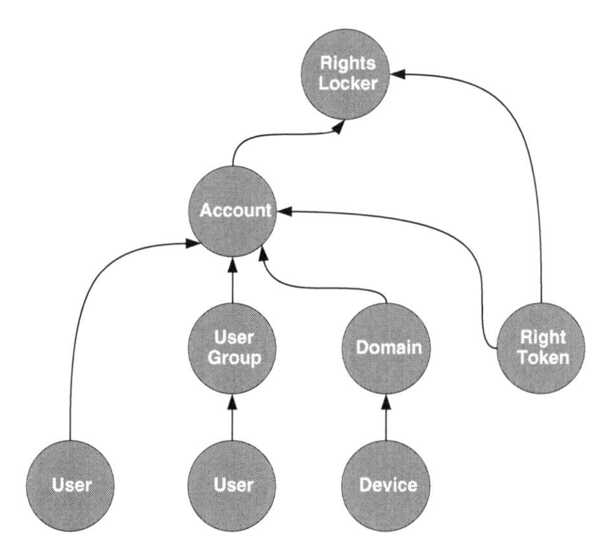

Fig. 11.9 DECE's organization

- Content owners are ensured that their content will be delivered with a certain level of security across any device. Thus, they may be keen to electronically release more valuable content.

Figure 11.9 shows the relations between the different elements of UV: the Account, the User, the User Group, the Device, the Domain, the Rights Locker, and the Rights Token. The linchpin of UV is its Account concept. An Account manages Users, User Groups, and Domains. The User Group encompasses all Users of a same household. A Domain represents the set of devices belonging to the same household. A User only interacts through her Account. For instance, a User interacts with her Account to add or remove a device from her Domain. The Rights Locker manages all Accounts. When a User purchases a piece of content, the retailer registers a Rights Token with the Rights Locker under the User's Account. The Rights Token describes the granted usage rights. The Rights Token is DRM-agnostic.

The obvious similarity with Coral (Sect. 11.3.1) is legitimate. In fact, many members of the DECE also participated in the Coral initiative. The main evolution is the concept of a common container. DECE defines four profiles of content: a portable device profile, an SD profile, a HD profile, and an ISO profile for DVD burning. DECE supports file download and streaming. Profiles ensure interoperability at the content protection layer. Using several profiles allows us to adapt the content to the capabilities of the receiving target device.

All DRM systems participating in the UV ecosystem will support the common container. For each piece of content, the common container holds the following:

- one or several URLs pointing to the native DRM license servers;
- the essence scrambled by a common bulk encryption;

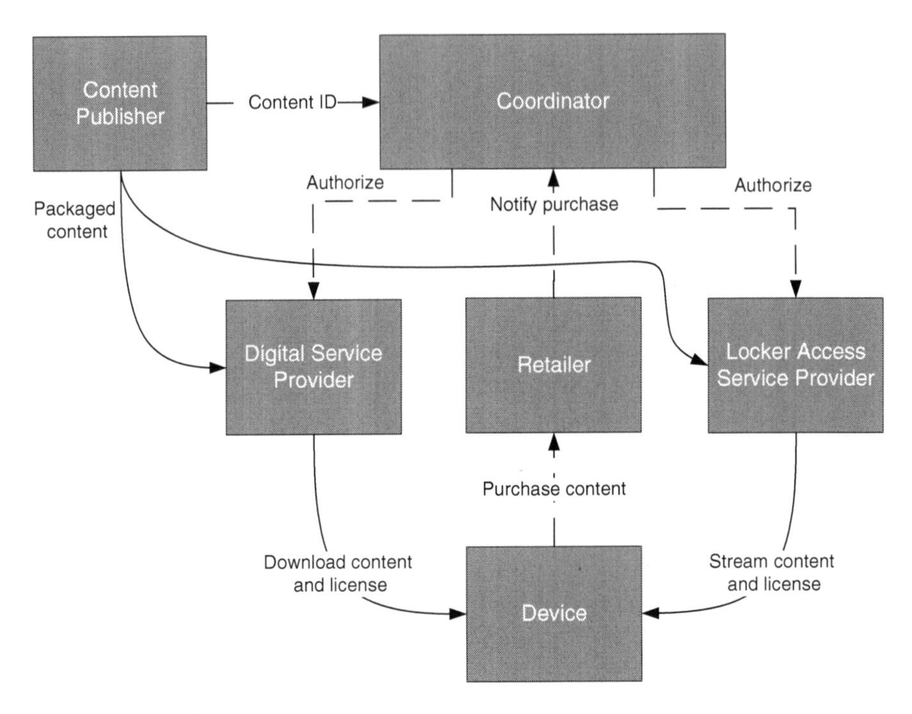

Fig. 11.10 DECE's roles

- a universal identifier;
- a data structure listing the scrambled sectors and a pointer to the corresponding decryption key; and
- a data structure holding native DRM licenses.

DECE defines a set of roles (Fig. 11.10):

- The coordinator ensures the interoperability of the system. It manages the Rights Locker, the Accounts, and the Domain management. The coordinator is the principal that authorizes content delivery and domain operations. DECE has already appointed company Neustar as the coordinator [337].
- Digital Service Providers manage the delivery of content when authorized by the coordinator. They generate the native licenses and the DRM native domain operations using the Rights Token provided by the Rights Locker.
- Locker Access Service Providers deliver streamed content to devices that are not part of a domain. The access is granted to a User or to an Account. They generate the native licenses using the Rights Token provided by the Rights Locker.
- Retailers sell the content to consumers. They notify the Coordinator of every purchase using the Rights Tokens.
- Content Publishers package the content and place the offer. Once the content published, it is available to Retailers, Digital Service Providers, and Locker

Access Service Providers. Content Publishers provide metadata for identification to the coordinator.
- Devices play back DECE content using one of the supported DRM systems. Currently, the supported DRM systems are Adobe Flash Access, Marlin, Microsoft PlayReady, OMA, and Widevine [338].

11.3.4 KeyChest

KeyChest is an initiative from the Walt Disney Company. It was first announced in 2009 [339]. KeyChest is an instantiation of the Interoperable Rights Locker. Like UltraViolet, it promises many advantages to consumers. KeyChest supports sell-through, rental, and subscription models and was designed to incorporate new models as they get promoted and adopted. KeyChest defines a media format container called SingleFile, whose use is optional and targeted at facilitating the digital supply chain.

The behavior of KeyChest is similar to that described in Sect. 11.2.5, Interoperable Rights Locker. Therefore, we will not explain its technical details. We will instead present two use cases to illustrate some benefits of digital rights lockers (DRLs) to consumers and distributors.

The first use case, illustrated in Fig. 11.11, demonstrates how an e-sell through transaction can work across several content providers. Let us assume that Alice has three devices: a computer, an IP Set Top Box, and a mobile device. The computer uses a software-based player using DRM 1. The STB and mobile phone are delivered by a network operator called Provider 2; The STB uses DRM 2, whereas the mobile phone uses OMA (marked as DRM 3 in the figure). The three devices are registered with KeyChest DRL under the name of Alice. Similarly, Provider 1 and Provider 2 are participants in KeyChest.

Alice purchases a movie from Provider 1 as an e-sell through transaction. Provider 1 records Alice's transaction by informing the KeyChest DRL that she purchased this movie. Once the transaction is registered, Provider 1 delivers the movie protected by DRM 1, and as expected, Alice can watch the movie from her computer. If Alice now wants to watch the movie on a TV set, the STB contacts Provider 2 who in turn queries KeyChest for Alice's entitlements for this movie. As Alice purchased the movie as an EST, the KeyChest DRL authorizes Provider 2 to deliver the movie to her STB and mobile phone. Provider 2 can stream the movie to Alice's STB using DRM 2 and to her mobile phone using OMA.

The second use case, illustrated in Fig. 11.12, is the digital copy scenario. Some content providers deliver DVDs or Blu-ray discs with the option to obtain the same movie as a digital file playable on computers or mobile devices. In other words, once Alice has purchased a DVD, she should be able to watch the same movie on her STB or mobile phone, although they are not equipped with a DVD drive.

In most implementations, the DVD or the Blu-ray disc contains a redemption code that is unique to the movie and to the instance, i.e., two DVDs of the same

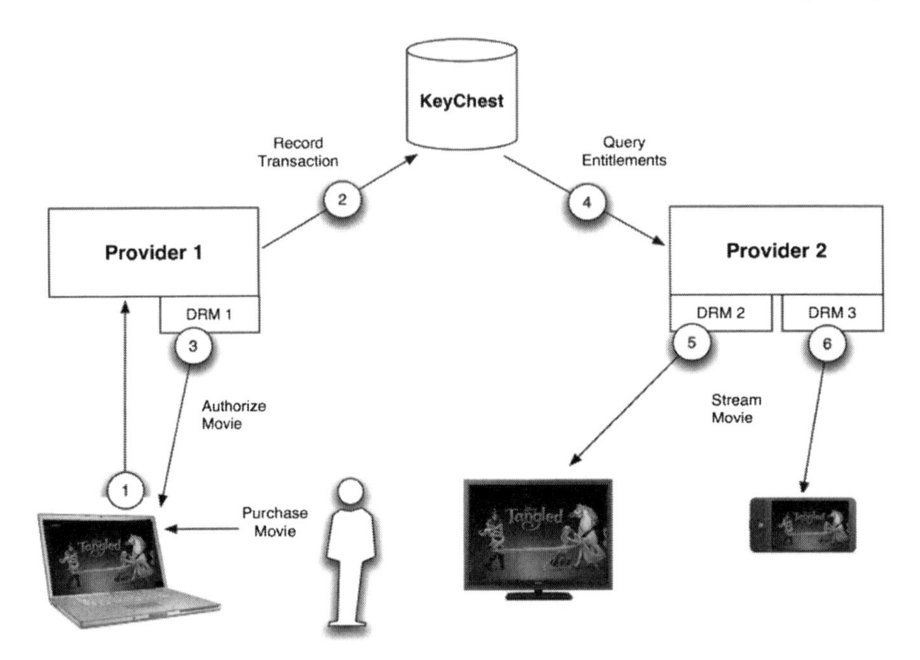

Fig. 11.11 KeyChest: cross provider use case (diagrams and images © Disney)

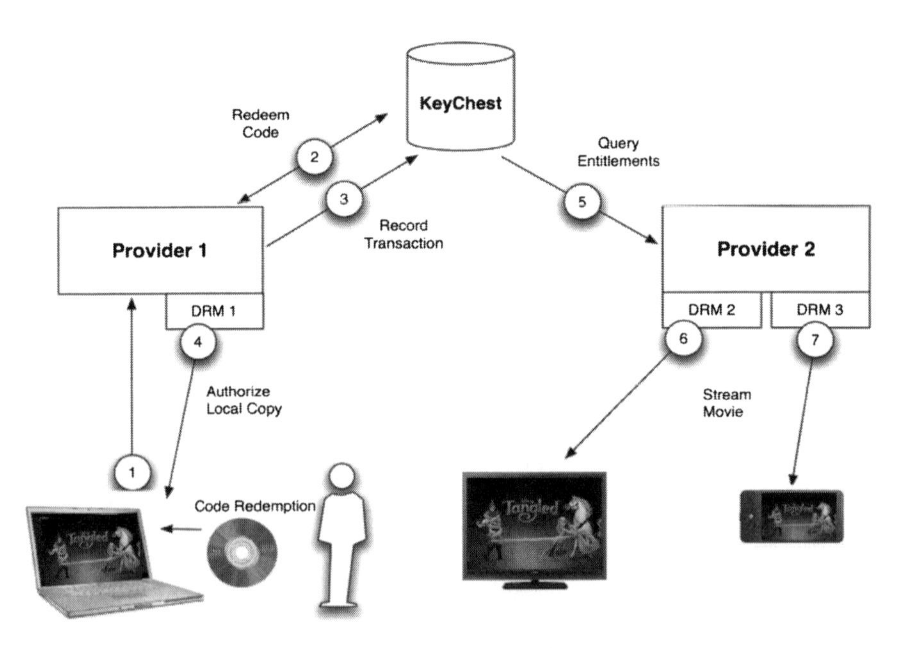

Fig. 11.12 KeyChest digital copy use case (diagrams and images © Disney)

movie have different redemption codes. When Alice inserts the newly purchased DVD for the first time in her computer, an application is launched and collects the redemption code and transfers it to Provider 1. Provider 1 checks with KeyChest DRL whether the redemption code is valid and has not yet been used. If this is the case, Provider 1 records the transaction by informing the KeyChest DRL that Alice redeemed the movie's redemption code. Provider 1 can then deliver a digital copy protected by DRM 1 to a computer. Since the transaction was recorded in the KeyChest DRL, Alice will be also able to watch the movie on her TV set and mobile phone as into the previous use case.

Disney's newly announced initiative, Disney Studio All Access, will use KeyChest at its core [340]. It has also been announced that KeyChest may be interoperable with UltraViolet.

11.3.5 Digital Media Project

The Digital Media Project (DMP), led by Leonardo Chariglione,[8] was founded in December 2003. Its goal is "to promote continuing successful development, deployment and use of Digital Media that respect the rights of creators and rights holders to exploit their works, the wish of end users to fully enjoy the benefits of Digital Media and the interests of various value-chain players to provide products and services" [341]. Currently, DMP has 21 members, among which are the École Polytechnique Fédérale de Lausanne, Telecom Italia, the University of Tokyo, Matsushita, Mitsubishi, BBC, and Telefonica. DMP developed the specifications of the Interoperable DRM Platform (IDP) and Chillout [342], its open-source implementation.

DMP issued three successive versions of its IDP. DMP published the first version in April 2005 and the second version in February 2006. The current version is IDP v3.3. To start, DMP compiled the requirements of many stakeholders such as civil rights associations, individuals, collective management societies, producers, FTA broadcasters, telecommunication operators, and consumer electronics manufacturers. It is interesting to highlight the absence of the main studios, the IT industry, and any CAS/DRM technology provider. Nevertheless, the result of this compilation is an impressive set of requirements and definitions of value chains. This has allowed to identify common primitive functions used by these value chains. The goal of IDP is to standardize these primitive functions and offer an interoperability framework.

IDP has defined a set of devices used by the different value chains [343]. Devices are either a combination of hardware and software or only a piece of software. In fact, devices are roles (or equivalent to the notion of Octopus nodes; see Sect. 7.6) (Fig. 11.13).

---

[8] Leonardo Chariglione is the founder of the MPEG standard and the SDMI.

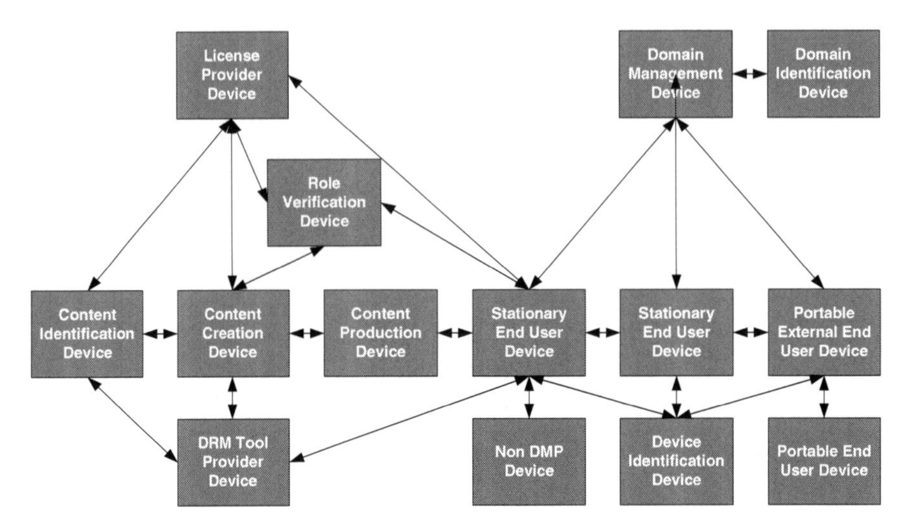

Fig. 11.13 Architecture of DMP's IDP

With IDP, content is a combination of resources. To create IDP content, the content creation device packages together resources, a license, DRM tools, and their respective identifiers. It receives the identifiers of resources, the license, DRM tools, and the created content itself from a content identification device managed by a registration authority. Content with an identifier is referred to as governed content. The license may be bundled with the content or delivered separately. Nevertheless, if the content creator does not want to impose any constraints on the consumption of her work, there will be no associated license. Similarly, DRM tools can be packaged together with the governed content. The created governed content is delivered to the content provider device.

If the governed content requires a license and DRM tools and if they are not bundled with the governed content, then the license (respectively DRM tools) is (are) delivered to the license provider device (respectively DRM tools provider).

IDP discriminates between fixed devices, called stationary end user devices, and mobile devices, called portable end user devices. A stationary end user device receives content from the content provider device. If the license is not bundled with the content, then the stationary end user device requests the license from the license provider device. If access to content needs DRM tools that were not bundled, and if the stationary end user device does not yet have these DRM tools, then it retrieves them from the DRM tool provider device. According to the usage rights granted by the license, the stationary end user device may consume the content, transform it, or transfer it to another stationary end user device or a portable end user device via a portable external device.

A portable external device receives content from a stationary end user device. If the license was not bundled with the content, then the portable external device requests it from the license provider device. According to the usage rights granted by the license, the portable external device may transform the content and transfer it to a portable end user device.

In the IDP model, portable end user devices do not have broadband or network access. Thus, they only receive bundled content from portable external devices. According to the usage rights granted by the bundled license, the portable end user device may consume the content.

We will have a brief look at DMP's DRM tools model as illustrated by Fig. 11.14. An end user device has several elements:

- A number of rendering pipelines which end up with decoders. In this example, there are two rendering pipelines. The number of decoders is device-dependent.
- A packager pipeline which allows us to export content to another end user device. Portable end user devices do not have such pipelines.
- A DRM processor which is a module that executes DRM-related functions in a trusted environment.
- Secure storage.
- A number of control points inside the pipelines; the DRM processor controls data in transfer at these points. In Fig. 11.14 the circles represent control points.

When an end-user device receives packaged content from a content provider device or another end user device, it parses the information. Resources to be rendered are pushed into corresponding rendering pipelines. Resources to be transformed and transferred to other end user devices are pushed into the packager pipeline. Once packaged, content is stored before being exported. Other information, such as licenses, DRM tools, or DRM information, is forwarded to the DRM processor.

If a DRM tool is missing, the DRM processor requests it from the DRM tool provider. When necessary, it instantiates a DRM tool and instructs it to operate on a given control point. For instance, it could require an AES decryption tool before storing data in Buffer 1 and a watermark embedder to operate before decoder 1. When the DRM processor does not need any more of these DRM tools, it stores all of them in its secure storage.

The trust management of DMP uses two types of authorities or agencies. Certification authorities check that devices comply with the specifications. Content creation devices, stationary end user devices, portable end user devices, and domain management devices need to be certified. The certification authority should enforce the C&R regime (which DMP has not yet defined and which would probably depend on the implemented value chain). Furthermore, DRM tools are also certified. Registration with authorities is mandatory for licenses, DRM tools, content creation devices, stationary end user devices, portable end user devices, and domain management devices. Only certified devices can be registered, i.e., receive an identifier.

DMP's interoperability relies on three elements: a standardized framework IDP, the use of a common certification authority and registration authority, and a way to download and execute DRM tools (plug-in-based interoperability). After 7 years of work, and three versions of specifications, there is still no commercial

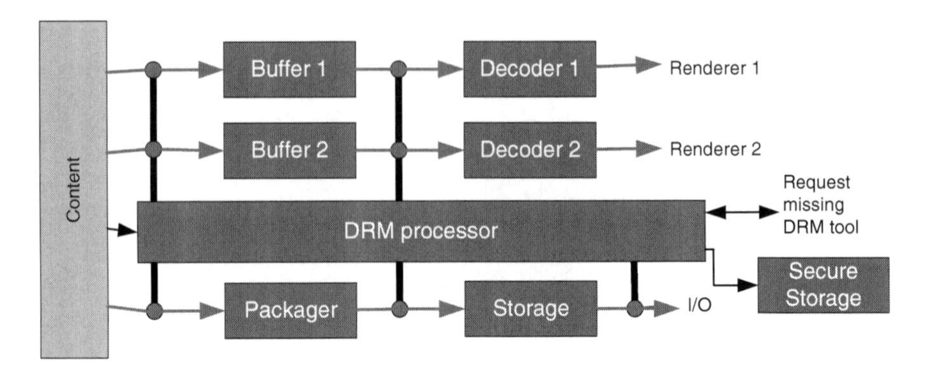

Fig. 11.14 DRM tool model

deployment of DMP-compliant DRM services. At the time of writing, DMP was still active.

11.4 Specific Problems

11.4.1 Transcoding of Encryption

Transcoding of scrambled content may seem a trivial problem. It is a complex problem [211]. The definition of a scrambling algorithm is not limited to the choice of an algorithm. We will describe in more detail the selection of a scrambling process.

Of course, the first choice is the selection of the cryptographic algorithm. It is either a stream cipher or a block cipher. In the case of block ciphers, it is mandatory to define the cipher mode, such as CBC or CTR. The padding conditions and the IV are other needed information. Applications may influence the decisions. For instance, if content requires persistent storage then encryption should occur before packetization. Scrambling has to take into account the content format, the transport mode, and the network.

Now let us focus on the problems inherent in interoperability. The first problem is the layer of scrambling in the Open Systems Interconnection (OSI) layer model. Not all scrambling methods operate at the same layer. Currently, solutions using the link layer, the network layer, the transport layer, and the presentation layer coexist. For instance, DTCP operates on the link layer; IPSec operates on the network layer; and ISMACryp and Microsoft operate on the presentation layer. The trusted agent may have to perform a translation of the scrambled layers. In this case, the translation is even more complex (Fig. 11.15).

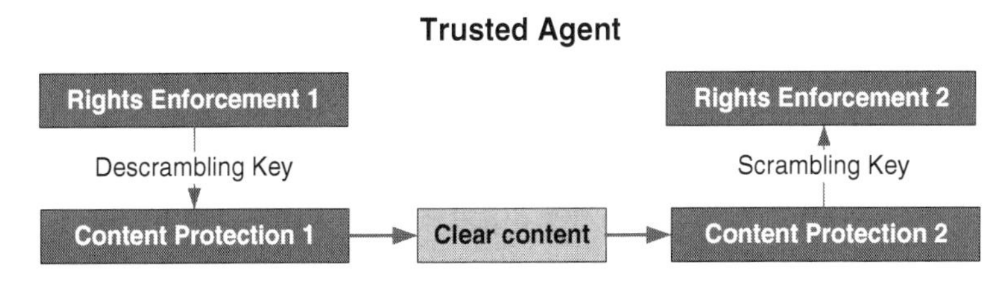

Fig. 11.15 Conversion of bulk encryption

11.4.2 Usage Rights Translation

The translation of usage rights between two DRM systems may become a complex issue for the following scenarios:

- REL differences;
- differences in usage rights; and
- more than two DRMs.

There is not one unique REL in the market. Several "public" languages are available and competing with each other. ODRL is the language chosen by the e-book forum and by SUN's OpenDReam. XrML is the language chosen by OMA and Microsoft. Furthermore, there is a plethora of proprietary languages. As with natural languages, translation is a difficult task.

The typical architecture for translating usage rights is a three-step process as illustrated in Fig. 11.16. We denote by REL_i the Rights Expression Language used by DRM_i.

The first step is to parse the usage rights expressed in REL_1. The parser interprets the usage rights of the incoming license. In the case of "publicly" available languages, such as ODRL, open-source references of the parser are available. In the case of proprietary languages, the parser must be implemented using this proprietary specification or is provided by the owner of the language.

The second step is the translation of usage rights. It maps each possible usage right from DRM_1 to a supported usage right from DRM_2. This mapping may be rather complex. Sometimes, the conversion is impossible. In that case, the rights converter issues an error message.

The final step encodes the mapped usage right in the REL of DRM_2. This step is straightforward. As for the parser of REL_1, it may use an open-source reference implementation or proprietary modules.

Designing the rights converter may be a complex task. We will start by addressing the "simple" problem of conversion from DRM_1 to DRM_2. In a second part, we will address the problem of several DRM systems and especially the case of chained translations.

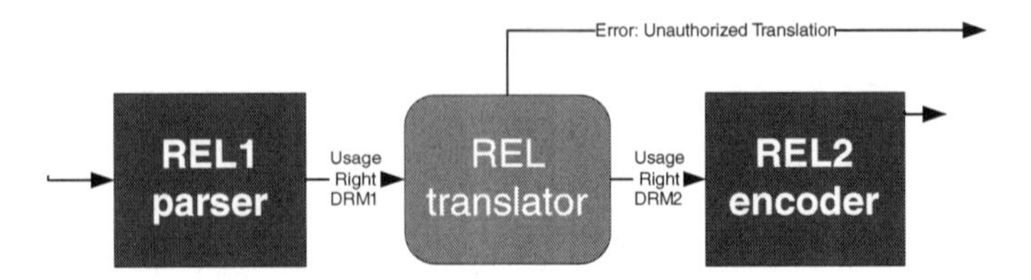

Fig. 11.16 Conversion of usage rights

Several parameters influence the translation. The first parameter is the expressivity of the REL used by each DRM. Four cases are possible, as illustrated in Fig. 11.17.

(a) REL_1 and REL_2 describe the same space of usage rights. In this case, the translation is straightforward. It is a bijection function. Each usage right expressed by REL_1 has one unique possible translation in REL_2.
(b) REL_1 is less expressive than REL_2. This case is a subset version of case a. REL_2 has a subset of usage rights that is equivalent to the space of REL_1. The translation will be a bijection function from the space of REL_1 to this subset of REL_2.
(c) REL_1 is more expressive than REL_2. Thus, only a subset of usage rights from REL_1 has a unique equivalent in DRM_2. This subset will be mapped one-to-one. The usage rights that are not in this common subset have three potential translation policies:

- The usage right from REL_1 is mapped to a less restrictive one (from the point of view of the consumer). The consumer will gain some perceived value.
- The usage right from REL_1 is mapped to a more restrictive one (from the point of view of the consumer). The consumer will lose some perceived value.
- The usage right from REL_1 cannot be mapped to any usage right from REL_2. In this case, the translation is not possible.

It may seem irrational to consider the last two cases, since they necessarily frustrate consumers. Nevertheless, several reasons may lead to these cases:

– Business rationale: The translation into more beneficial usage rights may destroy the logic of a commercial offer. An example will clarify the problem. Let us assume that a merchant proposes one piece of content with two usage rights of REL_1: R_{11} and R_{12}. R_{11} offers more perceived value for the consumers than R_{12}. Thus, the piece of content is sold at price x with usage right R_{11}, and at price y with usage right R_{12}. Of course, due to the perceived value, $y < x$. During translation, R_{11} has a direct equivalent usage right R_{21} in REL_2. Unfortunately, R_{12} has no direct equivalent in REL_2 and R_{21} is its "nearest" approximation. The mapping of R_{12} to R_{21} would kill the commercial offer linked to R_{11}. Soon, consumers would understand that content at the lower price

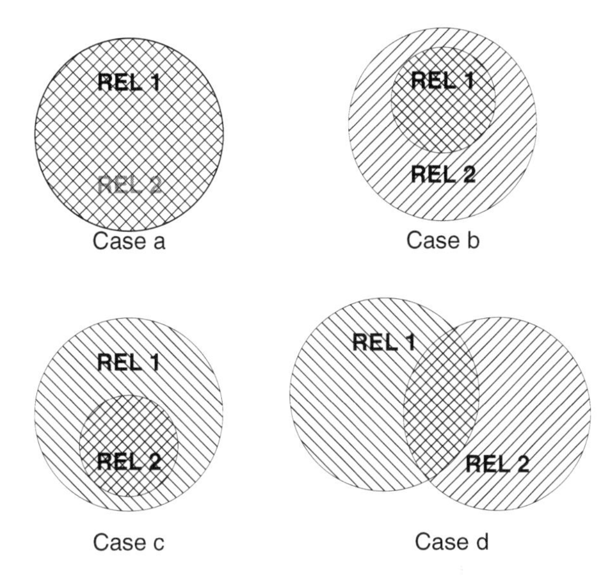

Fig. 11.17 Four REL translation cases

had the same value as the more expensive one if viewed with DRM_2. Thus, R_{12} should be mapped at least to a less favorable usage right than R_{21}.

- Contractual: Content providers define the conditions under which a piece of content can be distributed. The contract may introduce some strict constraints. Let us consider an example. REL_1 defines usage rights that specify the video quality of a copy. REL_2 only has a usage right that allows one copy, regardless of its video quality. A content provider may deliver HD content to the distributor. The content provider restricts copy authorization of HD content to downgraded content, i.e., with SD resolution. REL_1 can express this restriction. REL_2 cannot express it. Due to the contractual agreement, the translation from REL_1 to REL_2 will map the copy of REL_1 to "copy never" of REL_2 in the case of HD content.
- Confidentiality: This mainly relates to enterprise DRM. In this case, confidentiality is often adamant. To enforce confidentiality, translation will opt for more restrictive usage rights.

(d) REL_1 and REL_2 have different spaces of usage rights. Only a subset of these usage rights is common. The usage rights that belong to this common subset will have a one-to-one straightforward mapping. For the usage rights of REL_1 that are not part of the common subset, the translation policy will be similar to that of case c. It may be easier due to the non-overlapping usage rights of REL_2.

The problem becomes more complex if there are more than two interoperating DRM systems. If all the DRM systems are as in case (a), i.e., they use the same space of usage rights, then there is no problem. Else, a new problem arises: the loss of rights. To illustrate the problem, we will use three devices, D_1, D_2, and D_3. D_1 and D_3 use DRM_1 with REL_1, whereas D_2 uses DRM_2 with REL_2. We assume that DRM_1 and DRM_2 are as in case (c). Alice has one piece of content in device D_1. When transferring this protected content to device D_2, the translation of usage rights maps to more restrictive usage rights for DRM_2. When transferring content from device D_2 to device D_3, the translation defines the same restrictive usage rights for DRM_1. This is a loss of rights for the consumer. Of course, if Alice would have directly transferred the same piece of content from device D_1 to device D_3, she would not have lost any rights. Obviously, this situation is not acceptable to consumers. To deal with it, there are two strategies.

- Appending metadata to the converted license. The metadata would contain the usage rights before conversion. For instance, in the previous case, the metadata would contain the initial usage rights expressed in REL_1. DRM_2 does not use the metadata of the license. Nevertheless, if the license is converted back into DRM_1, the rights translator will use the usage rights of the metadata rather than the actual usage rights.

  The solution becomes difficult if the usage right is stateful, for example, view n times. A non-stateless usage right is modified during its lifetime. For instance, in this example, the license carries a viewing counter which is updated at each viewing. In this case, the rights enforcement layer of DRM_2 has to update both its own usage rights (expressed with REL_2) and the usage rights of the metadata (expressed with REL_1). This means that DRM_2 should partially implement the rights management layer of DRM_1.
- Using a central database that registers the purchased content. The database stores the purchase with the initial usage rights. A unique transaction number identifies each purchased item. Thus, before converting rights, the rights translator will query the central database with the unique identifier. The database will return the initial usage rights with the reference of the initial REL. The rights translator always applies the translation to the initial usage rights. This solution solves the problem of chaining DRMs and offers a consistent behavior to the consumer.

  If the usage rights require updating, the update is applied to the initial usage rights stored in the database (and of course also on the translated usage rights). This is the DRL strategy.

The translation of usage rights is a complex task. Part of this complexity comes from the strong restrictions due to business and contractual issues, which are always specific and need to be solved on a case-by-case basis.

11.5 Drawbacks

From a security point of view, interoperability has several drawbacks:

- Complexity: An interoperable DRM is more complex than an isolated DRM, which means more potential vulnerabilities either in the design or implementation. In design, interoperability means more use cases; hence, the probability that the designers forgot a use case during analysis increases with complexity. An attacker may find a use case that the designer did not foresee, and that gives advantage to the attacker. In implementation, complexity means more potential bugs that can be doors to vulnerabilities. Full coverage for test becomes impossible, or too expensive. Complex systems inherently are more difficult to secure than simple ones.
- Weakening of the ecosystem: A system is no stronger than its weakest link. If one of the interoperating DRM systems is broken, then this breach will affect content protected by all the other DRM systems, even if they were not broken. Pirates will use interoperability to bring content under the hood of the broken DRM, and thus freely access it. In theory, revocation should cope with the broken DRM. Nevertheless, revoking devices is a difficult, commercial decision.
- Secure implementation: Although security by obscurity is not a sound foundation for secure systems, a designer should not facilitate the work of the attackers by publishing the details of implementation. Often, theoretical protocols are secure; nevertheless, actual implementations may be weak. If the attacker already knows the implemented protocols, she has some elements to understand the implementation. Interoperable DRM systems require common documentation for APIs. These documents are valuable sources of information to the attackers.
- Complex business operations: Plug-in and translation approaches require additional components to those of the architecture described in Sect. 11.2.3. An operator other than the merchant or the DRM technology provider may operate these components. Who will pay for them?

Interoperable DRM systems require more careful implementation than proprietary solutions. Nevertheless, the benefits to consumers overcome these drawbacks. The risks are worthwhile.

Chapter 12
Some Challenges/Goodies

12.1 Open-Source DRM

Security based on open source is opposed to that based on closed source. It is common knowledge that openly reviewed cryptography is superior to proprietary cryptography. Most current cryptographic standards are issued from a long, open public reviewing process. It is thus legitimate to challenge the security of proprietary DRM solutions.

Security of open-source software, such as OpenSSL [344], is reputedly stronger than that of closed-source implementations. The reason for this better reputation is that a larger number of developers analyze the source code, and hence the higher the chance is that somebody will detect implementation flaws, for instance, buffer overflows. Unfortunately, security exploits do not only come from implementation flaws.

Like other open-source solutions,[1] open-source DRM has two elements: a public specification and reference source code. Let us analyze related issues. A common misconception is that the specifications of a DRM system need to be secret. This assertion is incorrect. Many specifications of copy protection schemes are publicly available, and their security model does not assume that they will remain secret. For instance, OMA [239] and AACS [346] are publicly available. Modern copy protection systems tend to comply with Kerckhoffs's law: "The system must not require secrecy and can be stolen by the enemy without causing trouble". This is good practice because specifications of consumer electronics devices always leak out.

The problem relies now on the strict implementation of Kerckhoffs's law: the need to keep secrets unknown and inaccessible to the attackers. Let us suppose that the attacker has access to the source code of the implementation and controls the execution environment. This would typically be the case for an open-source DRM

---

[1] We use the official definition of open source, as given by the open-source initiative [345]. Open source requires, for instance, free distribution, including of source code, authorization for derived works, and no discrimination against any field of endeavor.

E. Diehl, *Securing Digital Video*, DOI: 10.1007/978-3-642-17345-5_12,

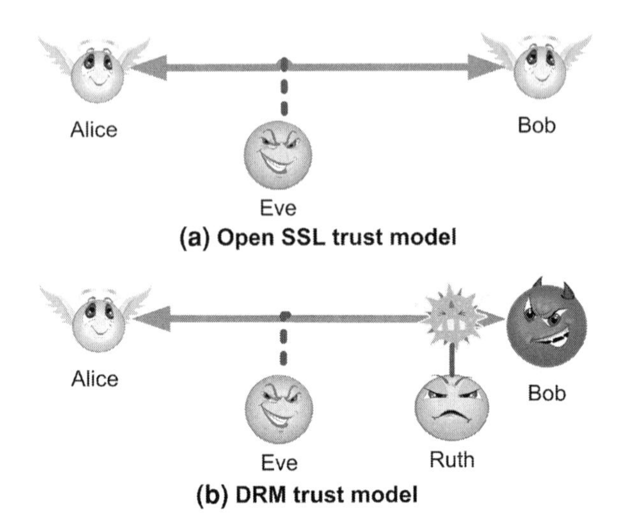

Fig. 12.1 Different trust models

running on a consumer's computer. With the source code, the attacker learns where the secret keys are stored, or at least when and where they are handled. Through the control of the environment, the attacker can extract these secret keys when and where they are available. Thus, the attacker can more easily hack the system.

If the source code is available, there is no possible way of efficient protection of the key in a software-based implementation, unless the attacker does not control the execution environment. Therefore, open-source DRM is not reliable for B2B and B2C. For instance, if a DRM requires a passphrase and the attacker has access to the source of the passphrase code, then she can easily eavesdrop on it, for instance, by overwriting the memory management [347].

People may object that if OpenSSL is more robust than proprietary implementations, then necessarily, open-source DRM should be more secure than proprietary solutions. Why is an implementation, such as OpenSSL, considered secure? As already mentioned, open-source code is reckoned more secure and stable because many developers have reviewed the source code. Therefore, it is expected that many bugs and weaknesses will be detected. This is probably true. Nevertheless, Open SSL is secure only under the trust model used for its initial design. SSL is a protocol which protects communication between two principals, Alice and Bob, who trust each other. The trust model assumes that Eve, the attacker, only controls the communication channel (Fig. 12.1a). Eve does not have access to the end-points. The attacker is not an insider. Knowing where OpenSSL stores the secret keys in the memory does not help Eve. Thus, Open SSL is efficient for secure communication. This trust model does not assume that Eve may partially control Bob's system. Were Eve to control Bob's computer, she could replace the OpenSSL implementation with her own implementation that would act like OpenSSL but would log all clear text. With the rise of malware, this may undermine the trust model.

The trust model of DRM is different [197]. The customer can be an attacker. Alice (the content owner) cannot trust Bob (the content user). Alice has no easy way to discriminate between an honest user and an unscrupulous one [348]. Using the cryptographic library of OpenSSL for implementation of DRM would reveal all secret keys to the attacker who controls the environment. Thus, such an approach is not suitable for DRM [172]. Even if Alice could trust Bob (for instance, in the case of enterprise DRM), she could not afford to trust Ruth, another attacker, not to partially control Bob's computer through malware (Fig. 12.1b). Indeed, a specially crafted malware could extract the secret keys. Similarly, if the DRM were to use a watermarking scheme for traceability, then in the case of an open-source implementation the attacker would have all the information to either wash the watermark or forge a watermark that would point to an innocent.

A valuable lesson is that it is fundamental to document the trust model of any design. When using a system for another usage or another context than the original ones, it is mandatory to verify that the trust model is still valid [192].

The second reason for not using open-source for DRM is the need to comply with a compliance and robustness regime. (See Sect. 3.8, Compliance and Robustness Rules.) This is not possible with open source. Open source does not provide any hook to enforce this regime. Anybody could write a variant that did not comply with C&R regime.

DReaM was an initiative of Sun Microsystems Laboratories. Its objective was to define an interoperability framework based on open standards [349] and offer an open-source-based DRM. It was in the category of translation-based interoperability. Although the specifications were completed and published, it was never deployed in the market. Sun labs did not define a C&R regime, nor propose a framework to certify implementations. Without such elements, major content providers will never offer content under this type of protection.

There may be one class of DRM for which an open-source implementation may be suitable: Consumer To Consumer (C2C). In C2C applications, we may assume that all participants are trustworthy. Thus, the open-source trust model holds. Furthermore, C2C DRM does not require any C&R regime. In that case, DRM based on open source makes sense. Nevertheless, this will not protect from an attacker such as Ruth, who uses a malware.

12.2 Clear Content DRM

One of the strongest assumptions in this book is that DRM-protected content is always scrambled. This is not always true. Recent approaches have provided an alternative path, where clear copyrighted content is distributed. Already, some merchant sites test sales of unscrambled mp3 songs. In December 2006, Yahoo Music was the first retailer to sell DRM-free mp3 songs [350]. The Digital Watermarking Alliance [351] and many observers applauded this approach. In February 2007, Steve Jobs pleaded in favor of DRM-free song distribution [234].

The songs are freely usable as long as they remain for personal use. With the advent of digital identifiers and mature identification technologies (Sect. 3.4, Fingerprinting), a new breed of DRM has appeared [352]. We will call it identifier-based DRM. Its main difference with a typical DRM is that its content is in the clear, rather than encrypted.

A first approach, based on watermarks, expects the rights enforcement to be voluntarily respected, and enforced by customers. It relies on people's goodwill. Nevertheless, a safeguard is provided. Every delivered song embeds a unique watermark that carries the identity of the customer. This is sometimes called transactional watermark or session watermark. Therefore, if the song is widely illegally distributed, there is a way to trace the originator of the unauthorized distribution. Some proponents have advocated that such an approach complies with fair use or private copy. Indeed the lack of scrambling does not introduce any technological barrier to the reuse of a work. Unfortunately, this approach may violate privacy issues [18]. A merchant or an authority may know the entire customer's viewing history. The use of a session watermark is an interesting challenge to privacy. The Center for Democracy & Technology proposed eight principles for enforcing privacy, such as avoiding embedding independently useful identifying information directly in a watermark; giving notice to end-users; controlling access to reading capability [353].

A more sophisticated approach proposes implementing rights enforcement even when using clear content. The approach is built on two hypotheses. The first hypothesis is that automatic identification of a piece of content is possible. There are two possible methods for this. The first method embeds a watermark in the essence of the content. The payload carries a unique identification of the title, and is applied before sending out the piece of content. This payload is common to every customer. The second method, proposed by a Philips research team [354], uses fingerprinting (Sect. 3.4, Fingerprinting). There is an accessible database which contains all the fingerprints of the monitored titles. In both cases, before playing back the piece of content, the DRM client attempts to identify it. It either extracts the watermark and directly retrieves the content's identity carried by the content's payload, or extracts fingerprints of the content and queries its identity in the reference database. Once the piece of content properly identified, the DRM client checks if the consumer has the corresponding license. If the consumer has the corresponding license, then the DRM client acts in compliance with the conditions defined in the license; else it does not play the content back. Obviously, this approach requires that only compliant readers play back the content. Any non-compliant player would skip the identification phase and directly play back the content, regardless of the presence of a valid license.

The two approaches are only valid if the objective is to keep honest people honest. None of these approaches is a solid barrier to piracy, but they depend on some human reactions:

- Honest people tend to remain honest if the perceived value of the content is acceptable. Using unscrambled distribution solves the problem of interoperability.

Fig. 12.2 Four-layer for identification-based DRM

Purchased content is available for all legal usage without restriction. It offers the same features as illegal content. Content perceived value is higher than that of scrambled and similar to that of pirated. Thus, customers may not look for unauthorized sources.

- Ease of use has value for many customers. Illegal content should be more complicated to retrieve than legal content. If, as previously mentioned, legal content offers the same commodities as its illegal counterparts do, many people will prefer the legal way.

In analogy with the four-layer model proposed in Sect. 4.5, we propose a four-layer model for this type of DRM. A content identification layer replaces the content protection layer (Fig. 12.2).

The content identification layer handles the actual identity of the content. This layer uses one of these three types of identification techniques: watermarking (Sect. 3.3), fingerprinting (Sect. 3.4), or digital hashing such as with MD5 (Sect. 3.2.2, Symmetric Cryptography). Thus, three cases occur:

- If the content identification layer uses watermarking, then the head end server embeds a watermark in the content prior to sending it to customers. The watermark's payload uniquely identifies the version of the content. It may be a unique ID of the content, or any other information that allows retrieving the identity of the piece of content, for instance, the emitter's name, the broadcast channel, and a time stamp. The content identification layer extracts the watermark and provides the data needed to identify the content.
- If the content identification layer uses fingerprinting, then the server extracts the fingerprints of the content to be sent to customers and stores them in the reference database of fingerprints. The client extracts the fingerprints from the received content, and compares them to the fingerprints in the reference database. The matching occurs either at the head end server or in the host; the former reduces the need for computing power and storage space while requiring more bandwidth.
- If the content identification layer uses cryptographic hash, then it computes the digital hash of the received content and compares it to hashes in a database. The database may be either a white list of contents, i.e., including all the authorized content, or a black list of contents, i.e., including all the unauthorized contents. The black list database may use the MD5 hash tags used by most P2P or DDL

Table 12.1 Watermark versus fingerprint versus digital hash

	Watermark	Fingerprint	Cryptographic hash
Modification of content	Invisible watermark	None	None
Process at the head end	Embed an invisible watermark in each piece of content before being emitted	Extract the fingerprints of the content to be identified. This is only done once for every version of the content	Calculate the cryptographic hash of the content to be identified. This is only done once for every version of the content
Process at the client side	Extract this watermark	Extract the fingerprints of the received content	Calculate the cryptographic hash of the received content
Accuracy	Rather good	Current systems have a reasonable recognition rate	There is no risk of error in identification but simple transformation defeats the system
Cost on client	Watermark extractor can for some technologies be implemented in consumer devices	Video fingerprint extractor requires computing power exceeding the capabilities of current consumer devices. It can be used by computer-based clients	Requiring reduced computing power. It can be used by computer-based clients and embedded devices

systems. The company Global File Registry proposes such a database [355]. Interestingly, Global File Registry uses an audio fingerprinting technology, provided by Audible Magic, for automatically populating its database, at least for illegal songs.

The choice of the identification technology affects the potential features and the exploitation mode of the system. Table 12.1 summarizes the main characteristics of each technology.

Digital hashing is less constraining in terms of computing power than the two other solutions. Unfortunately, digital hashing does not resist to any involuntary and voluntary attacks. By construction, the modification of one bit should result in a different cryptographic hash. The two other techniques are more robust to these types of attack. Therefore, if the content identification layer has to detect illegal content, in the case of digital hash, the database must collect an exhaustive list of all available unauthorized versions of the same work. If using watermarking technology, the content identification layer will spot any derivative of the initially watermarked work. Nevertheless, if the illegal content uses a non-watermarked source, then the content identification layer will fail. In the case of fingerprinting, the content identification layer will spot any version of a work.

The rights enforcement layer is rather simple on the server side. In the case of watermarking, it provides the payload that the content identification layer embeds. In the case of fingerprinting, it adds the extracted fingerprints to a reference database, if it is not yet present. In the case of digital hashing, the head end calculates the hash and stores it in a reference database.

On the client side, the rights enforcement layer has mainly two functions:

- Use the information provided by the content identification layer to definitively identify the content. It then forwards the information to the rights management layer.
- Enforce the command delivered by the rights management layer. There are three types of functions: Authorize the viewing, block the piece of content, or replace the piece of content by another one. This latter case could replace illegal content by a paying legal version or even propose a version with better quality.

It could be possible to enforce more complex usage rights, like the ones used by traditional DRMs. Nevertheless, in view of the brittle trust model of clear content DRM, it would be rather illusory.

Clear content DRM relies on a second strong assumption: ubiquitous compliant players. The content being in the clear, any player supporting the compression scheme can play it back. Thus, it is straightforward to circumvent. This scheme would only work if every manufacturers of video players, hardware-based or software-based, implemented this system. As there is no way to implement a hook IP (Sect. 3.8, Compliance and Robustness Rules), the best solution would be to have it made mandatory through national regulations. Unfortunately, the US broadcast flag (Sect. 6.1), which was built on a similar assumption, highlights the difficulty of such approach.

12.3 DRM and Game Theory

12.3.1 Introduction to Game Theory

Game theory studies the implications of rationality, self-interest, and equilibrium in decisions. Obviously, self-interest is one of the driving arguments of current DRMs. Content owners expect to maximize the return on investment of their production. Consumers expect to have access to compelling content at the lowest price. Pirates expect to get for free what is otherwise sold. The future of DRM will require finding a point of equilibrium that satisfies the interests of content owners, content distributors, and consumers. Are the decisions of all stakeholders driven by rationality? Probably not all past decisions were rationally driven. Thus, we may expect game theory to give some hints on defining the parameters for building DRM solutions that are acceptable and beneficial to all parties.

In game theory [356], player A has a set of possible actions $a = \{a_1, a_2 \dots a_n\}$ whereas player B has another set of possible actions $b = \{b_1, b_2, \dots, b_m\}$. Utility $u_A(a_i, b_j)$ is an evaluation of the benefit to player A if he chooses action a_i while player B chooses action b_j. The higher the utility, the more the benefit to the

Table 12.2 Table of utilities

Player A		Player B	
		b_1	b_2
	a_1	$U_A(a_1, b_1)$, $U_B(a_1, b_1)$	$U_A(a_1, b_2)$, $U_B(a_1, b_2)$
	a_2	$U_A(a_2, b_1)$, $U_B(a_2, b_1)$	$U_A(a_2, b_2)$, $U_B(a_2, b_2)$

player. A utility is negative when representing losses. The utility of every possible pair of actions has to be evaluated for each player. They are represented in tables such as that illustrated in Table 12.2. Each cell displays the utilities for both players if they choose the corresponding action.

An action a' dominates action a" for player A if all utilities for player A associated with action a' are better than those associated with action a" regardless of the actions of player B. In mathematical terms, it translates into $u_A(a', b_i) > u_A(a'', b_i) \forall a'' \in a, \forall i \in \{1, \ldots, m\}$. In that case, a rational player would never take the dominated action a". This action can thus be eliminated from the winning solutions.

The Nash equilibrium is at the heart of game theory and relies on the underlying assumption: all players play rationally. Thus, for any action by the opponent, a player will select the most rational action, i.e., the action with the highest utility for him. Naturally, rational players should end up at a Nash equilibrium. A Nash equilibrium is the mutual best response position. A game may have multiple Nash equilibriums. Unfortunately, not all games require rational players. In some games, players try to outguess each other (for instance, poker). In this case, each strategy will be balanced by a probabilistic distribution.

Nash equilibrium occurs if a solution is the best choice for both players. It is a mutual best response solution. In other words, (a*, b*) is not a Nash equilibrium if a* is not the best solution for player A if player B played b* and if b* is not the best solution for player B if player A played a*. In mathematical terms (a*, b*) is an instance of a Nash equilibrium if $u_A(a^*, b^*) \geq u_A(a', b^*)$, $\forall a' \in a$ and $u_B(a^*, b^*) \geq u_B(a^*, b')$, $\forall b' \in b$.

In the previous description, the game was static because we assumed that both players played simultaneously. In a dynamic game, the players play sequentially. First, player A selects an action a_1 from his set of feasible actions a. Then, player B selects an action b_1 from her set of feasible actions b. Once both players have played their turn, they evaluate their respective utilities. Of course, the decision of player B is influenced by the choice of a_1 by player A. Often, dynamic games are represented by a tree structure as illustrated in Fig. 12.3.

Often, the game is not limited to one round. The game is a set of consecutive rounds. Long-term gain influences the strategy of a rational player rather than just immediate gain. Such games are called repeated games.

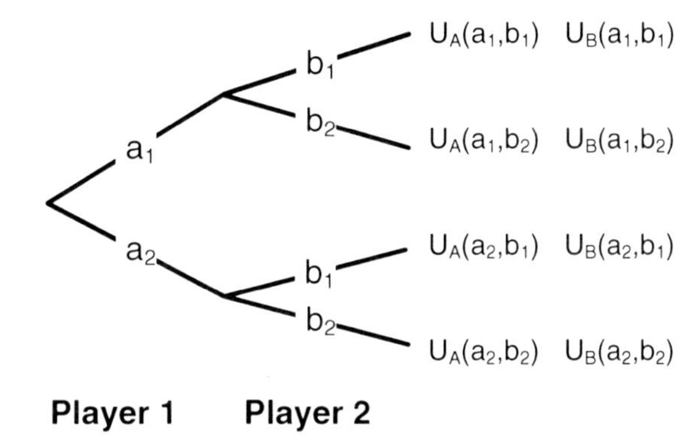

Fig. 12.3 Dynamic game

12.3.2 Current Research Topics

Gregory Heileman et al. [357] used the game theory to analyze the potential best strategy in a DRM field. This work is interesting, especially in the way that the conclusions are used. They use a simplified model of the DRM world by mapping it to two players: vendor and consumer. The strategy of the consumer is either to purchase the content from the vendor (*pay*) or to download it from free through an illegal site (*steal*). The strategy of the vendor is to sell the content in either a protected way (*drm*) or an unprotected way ($\overline{drm}$). Figure 12.4 illustrates the different utilities. The usual expectation of consumers is that $u_C(steal) \geq u_C(pay, \overline{drm}) \geq u_C(pay, drm)$, i.e., they may believe that free content has more perceived value than paid, non-protected content and, of course, DRM-protected content. Currently, the expectations of most vendors are that $u_V(pay, drm) \geq u_V(pay, \overline{drm}) \geq u_V(steal)$. We assume that DRM can provide efficient versioning or reduce piracy. The expectations of users and vendors are unfortunately opposed to each other.

Heileman introduces the concept of perfect DRM. A perfect DRM system is one such that content protected by it is not available for further copying. The option *steal* is unavailable to consumers. In this case, if $u_c(pay, drm)$ is not null, then theory shows that a rational vendor would never choose the $\overline{drm}$ strategy. Unfortunately, a perfect DRM is a utopian technology.

The study of a real case is more interesting. The critical parameter is the relation between $u_c(steal)$ and $u_c(pay, drm)$. If $u_C(steal) \geq u_C(pay, drm)$, then regardless of the price proposed by the vendor, there is a Nash equilibrium for strategy *steal*, i.e., the consumer is driven towards downloading rather than purchasing.[2] The only solution is to create a condition where $u_C(steal) \leq u_C(pay, drm)$. In that case, one Nash equilibrium appears for strategy (*pay*, *drm*). To reach this position, the vendor has to reduce utility $u_c(steal)$. There are two ways to do this: legal attacks and

[2] It should be noted that the notion of economic utility does not take into account any moral considerations. Moral conditions cannot be mapped to rational economic decisions.

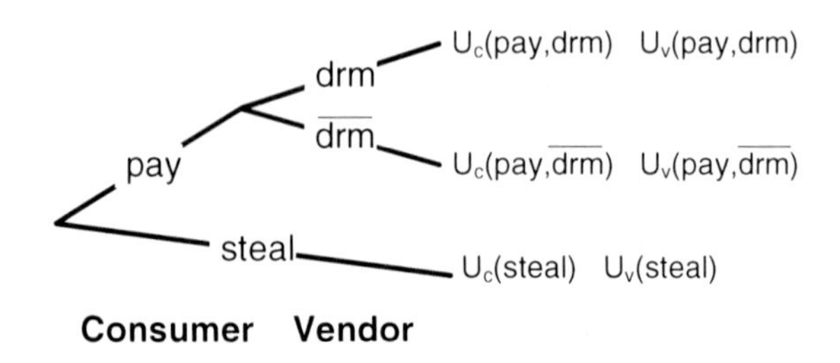

Fig. 12.4 Baseline DRM game

technical attacks. Legal attacks mean prosecuting illegal sharers. People being afraid of prosecution will drastically lower the value of $u_c(steal)$. Technological attacks spoil the downloaded content (bad quality, malware). Obviously, adding value to either $u_C(pay, drm)$ or $u_C(pay, \overline{drm})$ is also a solution to changing the ratio. Offering bonuses to legitimate consumers (*pay*), limited editions, and discounts on related events (such as concerts or premieres) are other possible solutions.

It is interesting to note that Heileman assumed that the stolen content would not be necessarily be offered as DRM-free ($\overline{drm}$), i.e., there is no strategy (*steal, drm*). This is not necessarily true in the real word. Sometimes it is possible to find, on unauthorized distribution networks, unauthorized content protected by some form of DRM (for instance, using a dedicated codec that has to been downloaded once paid for). This is especially true for early content, before theatrical release or just a few days after. Indeed, in that case, $u_C(steal, drm) \geq u_C(pay, drm)$ because the DRM-free version is not yet available. Normal consumers would not use such a strategy.

The subgames are extremely captivating. Subgames are the turns that occur once the baseline DRM game has completed. The consumer now has the content (either through *play* or through *steal*). The consumer now has two choices, *share*, and $\overline{share}$. The vendor has two possible strategies, *react*, and $\overline{react}$. Of course, we assume that the vendor is aware that its content is shared. In the case of large-scale piracy, this is a reasonable assumption. The actual vendor's reaction will depend on the subgame and action of the consumer. There are three possible subgames.

- The first subgame ($subgame_1$) corresponds to the case where the consumer purchased DRM-protected content. She may either share the purchased content with other people (*share*) or keep it ($\overline{share}$). The possible actions of distributor are either to sue (*react*) or not to sue ($\overline{react}$) the infringer.
- The second subgame corresponds to the case where the consumer purchased unprotected content. Here also, she has the choice to share or not. The vendor may adopt two strategies:
  - The vendor may grant the right to content only for personal usage. Thus, he can decide to sue the infringer.
  - He may decide to use the consumer as a distribution partner. Then his reaction will be to reward (*reward*) or not to reward ($\overline{reward}$) the consumer. This is called super-distribution.

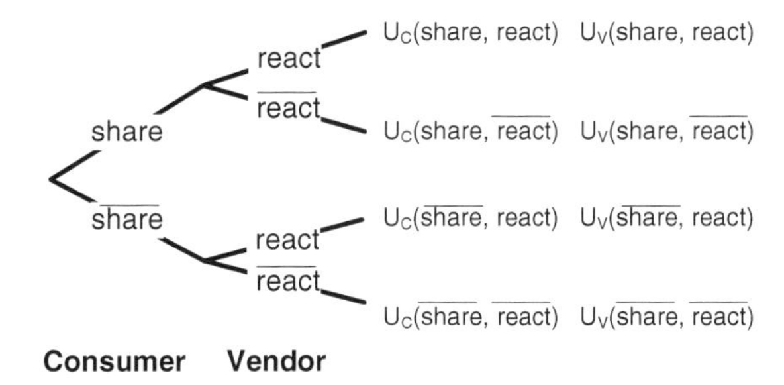

Fig. 12.5 Subgames for DRM

- The third subgame corresponds to the case of free riders on P2P networks. Free riders are P2P users that only download content and never upload content. In other words, free riders do not share content and resources on P2P networks [358] (Fig. 12.5).

Let us focus on the first subgame. To share DRM-protected content requires that at least one consumer breaks the corresponding protection. Thus, the vendor's action may be to sue the consumer. This is what RIAA routinely does when taking action against file sharers [359]. The vendor tries to set an example that the penalty on infringing consumers is severe. The stronger the punishment, the more will it be a deterrent to others [360]. Therefore, logically we may expect that $u_C(subgame_1, share, react) < < u_C(subgame_1, share, \overline{react})$. Unfortunately, legal action is costly. For most cases, the total cost of the legal action will exceed the loss incurred by the vendor. Thus, $u_V(subgame_1, share, action) < u_V(subgame_1, share, \overline{action})$. This is why RIAA goes mainly after big sharers and tries to reach faster settlements with infringers [361]. A settlement increases corresponding negative utilities $u_C(subgame_1, share, action)$, and $u_V(subgame_1, share, action)$.

The repeated games offer interesting perspectives. There are mainly two ways to analyze the repeated games. The first one is to always use the same content. In that case, we study the strategy of the vendor distributing a given piece of content to multiple consumers. Its outcome is to bias the game towards subgames 1 and 2, i.e., the subgames where the consumer purchases content. The vendor will not sue every player choosing subgame 3, due to economic constraints. The vendor will tolerate some leakage. The vendor will take legal action only after a certain amount of estimated loss. This strategy is called forgiving trigger. However, the vendor may choose many other strategies.

The second way to use repeated games is to always use the same vendor and customer but for different content. Here also, there are two potential strategies for the vendor. He may choose "no bad deeds got unpunished" or "no good deed got unrewarded". Currently, vendors choose the first strategy by applying it to a limited number of infringers. The cost of legal action is too expensive to generalize it.

Game theory helps us understand some current decisions of consumers and content owners. It would be interesting to model in a more realistic way the DRM arena. It would require at least three additional players: content owners, pirates who act as distributors of illegal contents but seek gain, and public authorities that decide to actively participate in the game, as is the case in France with HADOPI's graduated response.

Chapter 13
Conclusions

In Chap. 2, we built the case as to why digital video should be protected. Creating movies and TV shows is expensive. The cost of creating them is constantly increasing. The creators and investors expect a fair return on investment. Traditionally, copyright laws have provided an incentive to create with the promise of fair reward. Unfortunately, content digitalization has drastically facilitated piracy, online or in other forms. Although digitalization generates serious cost reduction for production and distribution, the incentive to invest proposed under the umbrella of copyright is in danger. The first aim of content protection may be to enforce fair reward by addressing digital piracy. Analog cultural goods often were offered in versioned packages. A song could be packaged in an album on 33 RPM LP records or audio tapes or as a single song on a 45 RPM LP record. Digital goods could easily offer such versioning if there were a way to create excludability. This is the second aim of video content protection. Furthermore, it is paramount for video content to be protected before its official release date: an early leak represents serious losses for content owners; hence, the B2B business needs efficient content protection solutions. Unfortunately, current solutions sometimes introduce severe restrictions on the use of digital goods. Consumers, being used to user-friendly analog goods, sometimes feel that these digital goods' restrictions are unfair and unnecessary. This resentment can be loudly expressed [362]. In retaliation, some consumers have deferred to alternate unauthorized sources of digital goods that do not impose usage limitations. This is increasingly putting at risk digital distribution.

Content owners have heard consumers' demands. The music industry is already changing its business models, looking for alternate methods to valorize music. The 2009 annual report of the International Federation of the Phonographic Industry (IFPI) has an exciting title: *New Business Models for a Changing Environment* [363]. Rather than focusing on piracy, as in previous years, the IFPI report has highlighted some new trends: Music access through subscription, ad-supported services, games, branding, and merchandising. DRM-free music is now common practice. However, the context of the music industry is simpler than that of the movie industry. For instance, it is difficult to believe that it would be possible to

E. Diehl, *Securing Digital Video*, DOI: 10.1007/978-3-642-17345-5_13,

produce a movie like James Cameron's *Avatar*, which cost over $300 million, with just the revenues of advertisement. DRM-free movies are not yet around the corner.

Nevertheless, content owners have proposed two evolutions. The first one is to develop new flexible business models. Offering a digital copy together with a physical medium is a first step. Already, some digital versions offer added value over the physical good in order to promote electronic sell-through (EST) [364]. Smart versioning, ranging from rental to EST, is the next frontier. Consumers will be able to choose a legitimate offer that will fit their willingness to pay for content. The second evolution is to relax the major constraint of DRM: device-based licenses. It took ten years from the initial work [365] on authorized domain-based content protection to see the concept of domain adopted by major DRM solutions and endorsed by content owners. Binding content to the consumer's domain should be acceptable to most users as it provides a certain flexibility. The large size of the consortiums designing such solutions (Coral, DVB-CPCM, UltraViolet) demonstrates the willingness from the movie industry, the distributors, and the technology providers to find a solution acceptable to consumers.

The next generation of video content protection will be interoperable. This is necessary for the economic health of the movie industry and the digital distribution value chain. One of the most efficient answers to digital piracy is to propose to consumers a product at the price and with the conditions they expect. Consumers want flexibility. The current motto is "anywhere and anytime". Video content protection will adapt to embrace this motto. It may take several trials to find the proper balance between acceptable user experience and strong security. Ultimately, video content protection will be transparent to honest people.

Appendix: Abbreviations—Acronyms

AAC	Advanced Audio Coding
AACS	Advanced Access Content System
AACS-LA	Advanced Access Content System Licensing Authority
ACTA	Anti-counterfeiting Trade Agreement
AES	Advanced Encryption Standard
AKE	Authenticated Key Exchange
API	Application Programming Interface
APS	Analog Protection System
ATR	Answer To Reset
ATSC	Advanced Television Systems Committee
B2B	Business-to-Business
B2C	Business-to-Consumer
BBS	Bulletin Board System
BD	Blu-ray Disc
BOPA	Break-Once Play-Always
BPDG	Broadcast Protection Discussion Group
C&R	Compliance and Robustness
C2C	Consumer-to-Consumer
CA	Certificate Authority
CA2DRM	Conditional Access to Digital Rights Management
CAS	Conditional Access System
CBC	Cipher Block Chaining
CC	Common Criteria
CCI	Copy Control Information
CD	Compact Disc
CD-A	Compact Disc Audio
CE	Consumer Electronics
CEK	Content Encryption Key
CGMS	Copy Generation Management System
CMLA	Content Management License Administrator
CPCM	Content Protection and Copy Management

E. Diehl, *Securing Digital Video*, DOI: 10.1007/978-3-642-17345-5,

CPPM	Copy Protection for Prerecorded Media
CPRM	Copy Protection for Recordable Media
CPSA	Content Protection System Architecture
CPTWG	Copy Protection Technical Working Group
CPU	Central Processing Unit
CRL	Certificate Revocation List
CSS	Content Scramble System
CTR	Counter mode
CW	Control Word
DCAS	Downloadable Conditional Access System
DCI	Digital Cinema Initiative
DCP	Digital Cinema Package
DCT	Discrete Cosine Transform
DDL	Direct Download
DECE	Digital Entertainment Content Ecosystem
DES	Data Encryption Standard
DFT	Discrete Fourier Transform
DLL	Dynamic Link Library
DMCA	Digital Millennium Copyright Act
DMP	Digital Media Project
DPA	Differential Power Attack
DPRL	Digital Property Rights Language
DRL	Digital Rights Locker
DRM	Digital Rights Management
DTA	Differential Timing Attack
DTCP	Digital Transmission Content Protection
DTLA	Digital Transmission Licensing Administrator
DVB	Digital Video Broadcast
DVB-CI	DVB Common Interface
DVB-CIT	DVB Common Interface Technical group
DVB-CPT	DVB Copy Protection Technical group
DVB-CSA	DVB Common Scrambling Algorithm
DVD	Digital Versatile Disc
DVD-CCA	DVD Copy Control Association
DVI	Digital Video Interface
DVR	Digital Video Recorder
DWT	Discrete Wavelet Transform
ECB	Electronic Code Book
ECC	Error Correcting Code or Elliptic Curve Cryptography
EC-DH	Elliptic Curve Diffie–Hellman
EC-DSA	Elliptic Curve Digital Signature Algorithm
ECM	Entitlement Control Message
EEPROM	Electrical Erasable Programmable Read-Only Memory
EFF	Electronic Frontier Foundation
EMM	Entitlement Management Message

EST	Electronic Sell-Through
EULA	End-User License Agreement
FCC	Federal Communications Commission
FIB	Focused Ion Beam
FTA	Free to Air
GPL	Gnu Public License
GSM	Global System for Mobile Communications
GUID	Generic User ID
HD	High Definition
HDCP	High-Bandwidth Digital Content Protection
HDMI	High-Definition Multimedia Interface
HMAC	Hash-Based Message Authentication Code
HSM	Hardware Security Module
IDP	Interoperable DRM Platform
IFPI	International Federation of the Phonographic Industry
IIS	Internet Information Services
IPMP	Intellectual Property Management and Protection
IRC	Internet Relay Chat
IRL	Interoperable Rights Locker
ISP	Internet Service Provider
IT	Information Technology
IV	Initialization Vector
KDM	Key Delivery Message
KID	Key Identifier
LAN	Local Area Network
MAC	Message Authentication Code
MKB	Media Key Block
MMS	Multimedia Message System
MPAA	Motion Picture Association of America
NEMO	Networked Environment for Media Orchestration
NIST	National Institute of Standards and Technology
NSA	National Security Agency
NSP	Network Service Provider
OCAP	Open Cable Application Platform
OCSP	Online Certificate Status Provider
ODRL	Open Digital Rights Language
OEM	Original Equipment Manufacturer
OMA	Open Mobile Alliance
OOB	Out-of-Band
OSI	Open Systems Interconnection
OTT	Over-the-Top
P2P	Peer-to-Peer
PCMCIA	Personal Computer Memory Card International Association
PGP	Pretty Good Privacy
PIFF	Portable Interoperable File Format

PIN	Personal Identity Number
PKCS	Public-Key Cryptography Standards
PKI	Public-Key Infrastructure
POD	Point of Deployment
PPV	Pay per View
PRNG	Pseudo-random Number Generator
PROM	Programmable Read-Only Memory
PUF	Physical Unclonable Function
PVR	Personal Video Recorder
QOS	Quality of Service
RAM	Random-Access Memory
RDD	Rights Data Dictionary
REL	Rights Expression Language
RIAA	Recording Industry Association of America
RPM LP	Revolutions per Minute Long-Playing
ROAP	Rights Object Acquisition Protocol
ROM	Read-Only Memory
RTT	Round-Trip Time
SAC	Secure Authenticated Channel
SACD	Société des Auteurs et Compositeurs Dramatiques
SD	Standard Definition
SDK	Software Development Kit
SDMI	Secure Digital Music Initiative
SIM	Subscriber Identity Module
SMPTE	Society of Motion Picture and Television Engineers
SMS	Subscriber Management System
SoC	System-on-Chip
SPA	Simple Power Attack
SPDC	Self-protecting Digital Content
SRM	System Renewability Message
SRTT	Secure Round-Trip Time
SSL	Secure Socket Layer
STA	Simple Timing Attack
STB	Set-Top Box
STTL	Secure Time to Live
TCG	Trusted Computing Group
TNT	Télévision Numérique Terrestre
TOC	Table of Contents
TPM	Trusted Platform Module
TTL	Time to Live
UCPS	Unified Content Protection System
UGC	User-Generated Content
UMTS	Universal Mobile Telecommunication System
URL	Universal Resource Locator
USB	Universal Serial Bus

uuid	Universally Unique Identifier
UV	Ultraviolet
VHS	Video Home System
VOD	Video-on-Demand
VPN	Virtual Private Network
WAN	Wide-Area Network
WHDI	Wireless High-Definition Interface
WIPO	World Intellectual Property Organization
WMDRM	Windows Media DRM
XrML	eXtended rights Markup Language

References

1. Berlind D (2006) FSF's Stallman pitches new definition for C.R.A.P. Between the Lines, Febr 2006. http://www.zdnet.com/blog/btl/fsfs-stallman-pitches-new-definition-for-crap/2582
2. EFF, The Corruptibles (2006) http://w2.eff.org/corrupt/
3. Rump N (2003) Definition, aspects, and overview. Digital rights management. Springer, Berlin, pp 3–15
4. Mossberg W (2002) Digital consumer takes up the fight against copyright plans in congress. Wall Street J
5. Horowitz P (2000) NARM poses baseline principles for e-commerce in music. Sounding board, pp 1–4. http://www.narm.com/news/sb/2000/0500/0500toc.htm
6. McCullagh D, Homsi M (2005) Leave DRM alone: a survey of legislative proposals relating to digital rights management technology and their problems. Mich State Law Rev 2005:317. http://heinonline.org/HOL/Page?handle=hein.journals/mslr2005&id=329&div=&collection=journals
7. Digital Millennium Copyright Act (1998) http://www.copyright.gov/legislation/dmca.pdf
8. Maillard T, Furon T (2004) Towards digital rights and exemptions management systems. Comput Law Secur Report 20:281–287. http://hal.archives-ouvertes.fr/inria-00083204/
9. A society for and by authors—SACD. http://www.sacd.fr/A-society-for-and-by-authors.750.0.html
10. Hardabud, Une histoire de la propriété intellectuelle dans le cyberspace. http://www.hardabud.com/alabordage/index.html
11. WIPO Copyright Treaty (1996) http://www.wipo.int/treaties/en/ip/wct/trtdocs_wo033.html
12. Contracting Parties: WIPO Copyright Treaty, WIPO. http://www.wipo.int/treaties/en/ShowResults.jsp?country_id=ALL&start_year=ANY&end_year=ANY&search_what=C&treaty_id=16
13. On the harmonisation of certain aspects of copyright and related rights in the information society (2001) http://ec.europa.eu/eur-lex/pri/en/oj/dat/2001/l_167/l_16720010622en00100019.pdf
14. Leyden J (2003) DVD Jon is free—official. The Register. http://www.theregister.co.uk/2003/01/07/dvd_jon_is_free_official/
15. Statement of the Librarian of Congress Relating to Section 1201 Rulemaking. US Copyright Office, July 2010. http://www.copyright.gov/1201/2010/Librarian-of-Congress-1201-Statement.html
16. Rosenoer J (1996) Cyberlaw: the law of the internet. Springer, New York

E. Diehl, *Securing Digital Video*, DOI: 10.1007/978-3-642-17345-5,

17. Gillespie T (2007) Wired Shut: copyright and the shape of digital culture. MIT Press, Cambridge
18. Burk D, ohen J (2000) Fair use infrastructure for copyright management systems. SSRN eLibrary, August 2000. http://papers.ssrn.com/sol3/papers.cfm?abstract_id=239731
19. Loi sur la propriété littéraire et artistique (1957) http://admi.net/jo/loi57-298.html
20. Loi relative aux droits d'auteur et aux droits des artistes-interprètes, des producteurs de phonogrammes et de vidéogrammes et des entreprises de communication audiovisuelle (1985)
21. Lemay JC, La taxe de copie privée. http://saceml.deepsound.net/copie_privee.html
22. Hugenholtz PB, Guibault L, van Geffen S (2003) The future of levies in a digital environment: final report. http://dare.uva.nl/record/128697
23. Bo D, Lévèque C-M, Marsiglia A, Lellouche R (2004) La piraterie de films: Motivations et pratiques des Internautes, May 2004
24. Varian HR (2005) Copying and copyright. J Econ Perspectives 19:121–138. http://www.ingentaconnect.com/content/aea/jep/2005/00000019/00000002/art00007
25. Dhamija R, Wallenberg F (2003) A framework for evaluating digital rights management proposals. 1st international mobile IPR workshop rights management of information products on the mobile internet, 2003, p 13
26. Messerschmitt D (1999) Some economics models. http://www.cs.berkeley.edu/~messer/netappc/Supplements/08-econm.pdf
27. Bird J (2009) Global entertainment market update 2009, Novemb 2009
28. Apple Unveils Higher Quality DRM-Free Music on the iTunes Store. www.apple.com, April 2007 available at http://www.apple.com/pr/library/2007/04/02itunes.html
29. Lesk M (2008) Digital rights management and individualized pricing. IEEE Secur Priv 6:76–79
30. Sobel L (2003) DRM as an enabler of business models: ISPs as digital retailers. Berkeley Technol Law J 18
31. About Creative Commons. http://creativecommons.org/about/
32. Tzu S (2002) The art of war. Dover Publications, Mineola
33. Biddle P, England P, Peinado M, Willman B (2002) The darknet and the future of content distribution. Lecture notes in computer science. Springer, Washington, DC, pp 55–176. http://www.cs.huji.ac.il/labs/danss/p2p/resources/the-darknet-and-the-future-of-content-distribution.pdf
34. Treverton G, Matthies C, Cunningham K, Goulka J, Ridgeway G, Wong A (2009) Film piracy, organized crime, and terrorism. The RAND Corporation. http://www.rand.org/pubs/monographs/2009/RAND_MG742.pdf
35. Singh D (2010) Real pirates of Bollywood: D-company. Daily News and Analysis, Sept 2010. http://www.dnaindia.com/mumbai/report_real-pirates-of-bollywood-d-company_1438589
36. Monnet B, Véry P (2010) Les nouveaux pirates de l'entreprise: Mafias et terrorisme. CNRS, Paris
37. Byers S, Cranor L, Korman D, McDaniel P, Cronin E (2003) Analysis of security vulnerabilities in the movie production and distribution process. Proceedings of the 3rd ACM workshop on digital rights management, Washington, DC, USA. ACM, pp 1–12. http://portal.acm.org/ft_gateway.cfm?id=947383&type=pdf&coll=GUIDE&dl=GUIDE&CFID=49793162&CFTOKEN=72317540
38. The pyramid of Internet Piracy (2005) http://www.mpaa.org/pyramid_of_piracy.pdf
39. Wallenstein A (2010) 'Hurt Locker' Studio Copyright Czar calls for end to DVD screener. PaidContent, Dec 2010. http://paidcontent.org/article/419-hurt-locker-studio-copyright-czar-calls-for-end-of-dvd-screeners/
40. Wang W (2004) Steal this file sharing book: what they won't tell you about file sharing. No Starch Press, San Francisco
41. Diehl E, Wolverine is a success … The blog of content security. http://eric-diehl.com/blog/?x=entry:entry090408-185426

42. Le Blond S, Legout A, Lefessant F, Dabbous W, Ali Kaafar M (2010) Spying the world from your laptop. 3rd USENIX workshop on large-scale exploit and emergent threats (LEET'10), San Jose, CA, USA, 2010. http://www.mpi-sws.org/~stevens/pubs/leet10.pdf
43. Michaud L (2008) Echange et piratage de contenus Nouvelles pratiques, nouveaux outils, IDATE, 2008
44. Megauploads—Rewards. http://www.megaupload.com/?c=rewards&setlang=en; US Department of Justice, Justice Department Charges Leaders of Megaupload with Widespread Online Copyright Infringement, January 2012. http://www.justice.gov/opa/pr/2012/January/12-crm-074.html
45. Baio A, Pirating the 2009 Oscars. Waxy. http://waxy.org/2009/01/pirating_the_2009_oscars/
46. Jardin X (2006) DVD pirates successfully plunder Academy Award screeners. Slate Mag, January 2006. http://www.slate.com/id/2134292/
47. GVU, VAP, and SAFE, Filmschutz vor und bei Kinostart, June 2008
48. Morris S (2010) New to apps: Torrent discussions with Torrent Tweet. BitTorrent Blog, Aug 2010. http://blog.bittorrent.com/2010/08/05/new-to-apps-social-commenting-with-torrent-tweet/
49. Eskicioglu A (2005) Multimedia content protection in digital distribution networks. http://www.sci.brooklyn.cuny.edu/~eskicioglu/Tutorial/Tutorial.pdf
50. Rovi, RipGuard: protecting DVD content owners from consumer piracy. http://www.rovicorp.com/products/content_producers/protect/ripguard.htm
51. Murphy K (2009) Comments of Paramount Pictures Corporation in response to the national broadband plan workshop on the role of content in the broadband ecosystem, Oct 2009. http://reporter.blogs.com/files/09-51-10-30-2009-viacom-inc._paramount-pictures-corporation-1-of-2-7020244125.pdf
52. Yager L (2010) Intellectual property: observations on efforts to quantify the economic effects of counterfeit and pirated goods, GAO, 2010. http://www.gao.gov/new.items/d10423.pdf
53. Diehl E (2010) Be our guest: Chris Carey. Technicolor Secur Newsl 5, July 2010. http://www.technicolor.com/en/hi/research-innovation/research-publications/security-newsletters/security-newsletter-16/be-our-guest
54. Al, What you get for being a law-abiding paying customer. The Blog of Record, Febr 2010. http://www.theblogofrecord.com/2010/02/23/pirate-dvd-vs-legal-dvd/
55. Diehl E (2008) Content protection: in this digital age, protecting your assets is essential. Broadcast Eng, March 2008. http://www.nxtbook.com/nxtbooks/penton/be0308/index.php?startpage=10
56. Bell DE, LaPadula LJ (1973) Secure computer systems: mathematical foundations. http://stinet.dtic.mil/oai/oai?&verb=getRecord&metadataPrefix=html&identifier=AD0770768
57. Benantar M (2010) Access control systems: security identity management and trust models. Springer, Heidelberg
58. Garfinkel S, Spafford G, Schwartz A (2003) Practical UNIX & internet security, 3rd edn. O'Reilly Media, Cambridge
59. McClure S, Scambray J, Kurtz G (2009) Hacking exposed 6: network security secrets & solutions. Osborne/McGraw-Hill, Emeryville
60. Lefebvre F, Arnold M (2006) Fingerprinting and filtering. Secur Newsl, Dec 2006. http://eric-diehl.com/newsletterEn.html
61. Walker K (2010) Making copyright work better online. Google Public Policy Blog, Dec 2010. http://googlepublicpolicy.blogspot.com/2010/12/making-copyright-work-better-online.html
62. Christin N, Weigend A, Chuang J (2005) Content availability, pollution and poisoning in file sharing peer-to-peer networks. Proceedings of the 6th ACM conference on electronic commerce, Vancouver, Canada

63. Shoup V (2008) A computational introduction to number theory and algebra. Cambridge University Press, Cambridge
64. Menezes A, Van Oorschot P, Vanstone S (1996) Handbook of applied cryptography. CRC Press, Boca Raton
65. Wolf P (2003) De l'authentification biométrique. Sécurité Informatique, Oct 2003. http://www.sg.cnrs.fr/FSD/securite-systemes/revues-pdf/num46.pdf
66. von Ahn L, Blum M, Langford J (2004) Telling humans and computers apart automatically. Commun ACM 47:56–60. http://portal.acm.org/ft_gateway.cfm?id=966390&type=html&coll=GUIDE&dl=GUIDE&CFID=109469816&CFTOKEN=32876405
67. Mori G, Malik J, Breaking a Visual CAPTCHA. http://www.cs.sfu.ca/~mori/research/gimpy/
68. Maetz Y, Onno S, Heen O (2009) Recall-A-Story, a story-telling graphical password. Mountain View
69. Kahn D (1976) The code-breakers. Macmillan, New York
70. Kerckhoffs A (1883) La cryptographie militaire. Journal des sciences militaires
71. Blaze M, Diffie W, Rivest RL, Schneier B, Shimomura T (1996) Minimal key lengths for symmetric ciphers to provide adequate commercial security. A report by an ad hoc group of cryptographers and computer scientists. http://stinet.dtic.mil/oai/oai?&verb=getRecord&metadataPrefix=html&identifier=ADA389646
72. Lenstra AK, Verheul ER (2001) Selecting cryptographic key sizes. J Cryptol 14:255–293. http://dx.doi.org/10.1007/s00145-001-0009-4
73. Loukides M, Gilmore J (1998) Cracking DES: secrets of encryption research, wiretap politics and chip design. O'Reilly Associates, Sebastopol
74. Daemen J, Rijmen V (2002) The design of Rijndael: AES—the advanced encryption standard. Springer, Berlin
75. Wang X, Yin YL, Yu H (2005) Finding collisions in the full SHA-1. http://citeseerx.ist.psu.edu/viewdoc/summary?doi=10.1.1.94.4261
76. Biham E, Chen R, Joux A, Carribault P, Lemuet C, Jalby W (2005) Collisions of SHA-0 and reduced SHA-1: advances in cryptology—Eurocrypt 2005. Springer, Aarhus
77. Diffie W, Hellman M (1976) New directions in cryptography. IEEE Trans Inf Theory 22:644–654
78. Rivest RL, Shamir A, Adleman L (1978) A method for obtaining digital signatures and public-key cryptosystems. Commun ACM 21:120–126
79. X.509: Information technology—Open Systems Interconnection—The Directory: Public-key and attribute certificate frameworks. http://www.itu.int/rec/T-REC-X.509/en
80. Public-Key Cryptography Standards (PKCS), RSA Laboratories. http://www.rsa.com/rsalabs/node.asp?id=2124
81. Cox I, Miller M, Bloom J, Fridrich J, Kalker T (2007) Digital watermarking and steganography. Morgan Kaufmann, San Francisco
82. Walker R, Mason A, Salmon RA (2000) Conditional access and watermarking in 'free to air' broadcast applications, Sept 2000
83. Moon TK (2005) Error correction coding: mathematical methods and algorithms. Wiley-Blackwell, Malden
84. Voloshynovskiy S, Pereira S, Pun T, Eggers JJ, Su JK (2001) Attacks on digital watermarks: classification, estimation based attacks, and benchmarks. Commun Mag IEEE 39:118–126
85. Petitcolas F (2009) Stirmark benchmark 4.0. Fabien Petitcolas personal website, June 2009. http://www.petitcolas.net/fabien/watermarking/stirmark/
86. Samtani R (2009) Ongoing innovation in digital watermarking. Computer 42:92–94. 10.1109/MC.2009.93
87. Kirstein M (2008) Beyond traditional DRM: moving to digital watermarking & fingerprinting in media. MultiMedia Intelligence

88. Reitmeier G, Lauck M (2007) Beyond the broadcast flag—content and signal protection approaches for broadcasters. http://www.nabanet.com/nabaweb/documents/agms/2008/NABA%20TC%20-%20Content%20Protection%20for%20Broadcasters.pdf
89. Wagner NR (1983) Fingerprinting. Proceedings of the 1983 Symposium on Security and Privacy, Oakland, CA, USA. IEEE Computer Society Press, pp 18–22. http://www.cs.utsa.edu/~wagner/pubs/finger/fing2.pdf
90. Huffstutter P (2003) How Hulk crushed the online pirate. Los Angeles Times, June 2003. http://articles.latimes.com/2003/jun/26/business/fi-hulk26
91. Furon T, Pérez-Freire L (2009) Worst case attacks against binary probabilistic traitor tracing codes. 1st IEEE international workshop on information forensics and security, WIFS 2009, IEEE, 2009, pp 56–60
92. Furon T, Pérez-Freire L (2009) EM decoding of Tardos traitor tracing codes. Proceedings of the 11th ACM workshop on multimedia and security, Princeton, New Jersey, USA, ACM, 2009, pp 99–106. http://portal.acm.org/citation.cfm?id=1597817.1597835
93. Carmine Caridi, Motion Picture Academy member who handed over his awards screeners for illegal duplication, ordered to pay $300,000 to Warner Bros. Entertainment Inc., Novemb 2004. http://www.timewarner.com/corp/newsroom/pr/0,20812,832500,00.html
94. Fighting audiovisual piracy: a good practice guide for the industry (2007) http://www.cnc.fr/Site/Template/T8.aspx?SELECTID=2531&ID=1661&t=1
95. Digital Cinema System Specification, Version 1.2, March 2008. http://www.dcimovies.com/DCIDigitalCinemaSystemSpecv1_2.pdf
96. Chupeau B, Massoudi A, Lefebvre F (2008) In-theater piracy: finding where the pirate was. Security, forensics, steganography, and watermarking of multimedia contents. SPIE, San Jose, p 68190T-10
97. E-City nabs pirates using Thomson Watermarking tech. Businessofcinema.com, Febr 2009. http://businessofcinema.com/news.php?newsid=12167
98. Sadun E (2007) Don't Torrent that song ..., The Unofficial Apple Weblog, May 2007. http://www.tuaw.com/2007/05/30/tuaw-tip-dont-torrent-that-song/2
99. Kundur D, Karthik K (2004) Video fingerprinting and encryption principles for digital rights management. Proceedings of the IEEE, vol 92
100. Le Buhan C, Rufle D, Stransky P, Verbesselt I (2007) Forensic watermarking architecture in a CAS/DRM environment. http://www.nagra.com/dtv/company/newsroom/document-center/
101. Eskicioglu A, Delp EJ (April 2001) An overview of multimedia content protection in consumer electronics devices. Signal Process Image Commun 16:681–699
102. Zhao J, A WWW service to embed and prove digital copyright watermarks. European conference on multimedia applications, services and techniques, pp 695–710
103. Mason A, Salmon RA, Devlin J (2000) User requirements for watermarking in broadcast applications. International broadcasting convention (IBC 2000), Amsterdam, The Netherlands, 2000
104. Piva A, Bartolini F, Barni M (2002) Managing copyright in open networks. IEEE Internet Computing, 2002, pp 18–26
105. Radio & TV Audience Measurement: NexTracker (2009). http://www.civolution.com/fileadmin/bestanden/PDF/nexguard/Leaflet_NexTracker_-_Audience_Measurement.pdf
106. Le watermarking: ça coule de source (2009). http://www.mediametrie.fr/innovation/pages/le-watermarking-ca-coule-de-source.php?p=7,112,78&page=85
107. TVaura for Live Broadcast Tracking (2010). http://www.tvaura.com/assets/pdf/TVaura_WP_VideoBroadcastTracking.pdf
108. Lefebvre F, Chupeau B, Massoudi A, Diehl E (2009) Image and video fingerprinting: forensic applications. Proceedings of the SPIE, San Jose, CA, USA, 2009, pp 725405–725405-9. http://spie.org/x648.html?product_id=806580

109. Cano P (2006) Content-based audio search: from fingerprinting to semantic audio retrieval. Pompeu Fabra University, 2006. http://citeseerx.ist.psu.edu/viewdoc/download?doi=10.1.1.75.738&rep=rep1&type=pdf
110. Massoudi A (2008) Combining fingerprinting and selective encryption for image and video protection. Université catholique de Louvain, 2008
111. Lefebvre F (2004) Message digest for photographic images and video contents. Université catholique de Louvain, 2004
112. Markantonakis K, Mayes K (2008) Smart cards, tokens, security and applications. Springer, Heidelberg
113. Moreno R (1976) Methods of data storage and data storage systems. US Patent 3,971,916, 27 July 1976
114. Guillou L, Histoire de la carte à puce du point de vue d'un cryptologue. Actes du 7eme Colloque sur l'Histoire de l'Informatique et des Transmissions, pp 126–155. http://www.aconit.org/histoire/colloques/colloque2004/guillou.pdf
115. ISO—International Organization for Standardization. ISO -7816-1:1987 Cartes d'identification—Cartes à circuit(s) intégré(s) à contacts. http://www.iso.org/iso/fr/iso_catalogue/catalogue_ics/catalogue_detail_ics.htm?csnumber=14732
116. ISO—International Organization for Standardization. ISO-IEC 10536—identification cards—contactless integrated circuit(s) cards—Close-coupled cards, April 2000
117. ISO—International Organization for Standardization. ISO-IEC 14443 Identification cards—contactless integrated circuit(s) cards—proximity cards, 2008
118. ISO—International Organization for Standardization. ISO-IEC 15693 Cartes d'identification—Cartes à circuit(s) intégré(s) sans contact—Cartes de voisinage, 2000
119. FIPS 140-2 security requirements for cryptographic modules, May 2001. http://csrc.nist.gov/publications/fips/fips140-2/fips1402.pdf
120. Abraham DG, Dolan GM, Double GP, Stevens JV (1991) Transaction security system. IBM Syst J 30:206–229. http://portal.acm.org/citation.cfm?id=103494.103495
121. IBM PCI Cryptographic Coprocessor, Dec 2008. http://www-03.ibm.com/security/cryptocards/pcicc/overview.shtml
122. NIST, Validated 140-1 and 140-2 cryptographic modules. http://csrc.nist.gov/groups/STM/cmvp/documents/140-1/140val-all.htm
123. Trusted Computing Group: TPM. https://www.trustedcomputinggroup.org/groups/tpm/
124. Arbaugh WA, Farber DJ, Smith JM (1997) A secure and reliable bootstrap architecture. Proceedings of the IEEE symposium on security and privacy, Oakland, CA, USA, pp 65–71
125. Anderson R (2003) Trusted computing FAQ TC/TCG/LaGrande/NGSCB/Longhorn/Palladium, August 2003. http://www.cl.cam.ac.uk/~rja14/tcpa-faq.html
126. The Silicon Zoo. http://www.siliconzoo.org/
127. Bar-El H, Choukri H, Naccache D, Tunstall M, Whelan C, Rehovot I (2006) The Sorcerer's Apprentice guide to fault attacks. Proc IEEE 94:370–382
128. Kocher PC (1996) Timing attacks on implementations of Diffie-Hellman, RSA, DSS, and other systems. Lecture notes in computer science, vol 1109, pp 104–113. http://citeseer.ist.psu.edu/kocher96timing.html
129. McGraw G (June 2002) On bricks and walls: why building secure software is hard. Comput Secur 21:229–238
130. Sander T, Tshudin C (1998) Protecting mobile agents against Malicious Hosts. Mobile agents and security, Vigan Giovanni, 1998, pp 44–60. http://www.springerlink.com/content/tp171en7td8uvchu/
131. van Oorschot PC (2003) Revisiting software protection. Proceedings of the 6th international conference on information security, Bristol, UK, 2003
132. Schneier B (2004) Secrets and lies: digital security in a networked world. Hungry Minds Inc., New York

133. Main A, van Oorschot PC (2003) Software protection and application security: understanding the battleground, Heverlee, Belgium, 2003. http://www.scs.carleton.ca/~paulv/papers/softprot8a.pdf
134. Courtay O (2006) DRM war: news from the front. Thomson Secur Newsl 1
135. Barak B, Goldreich O, Impagliazzo R, Rudich S, Sahai A, Vadhan SP, Yang K (2001) On the (im)possibility of obfuscating programs, advances in cryptology—Crypto 2001. Lecture notes in computer science, vol 2139, pp 1–18
136. Collberg CS, Thomborson C (2002) Watermarking, tamper-proofing, and obfuscation—tools for software protection. Trans Softw Eng 28:735–746
137. SoftICE, Wikipedia, the free encyclopedia. http://en.wikipedia.org/wiki/SoftICE
138. IDA Pro Disassembler—multi-processor, windows hosted disassembler and debugger. http://www.datarescue.com/idabase/
139. OllyDbg v1.10. http://www.ollydbg.de/
140. GDB: The GNU Project Debugger. http://www.gnu.org/software/gdb/
141. Quist D, Ames C (2008) Temporal reverse engineering. Black Hat Briefings, Las Vegas
142. Torrubia A, Mora FJ (2000) Information security in multiprocessor systems based on the X86 architecture. Comput Secur 19:559–563
143. Falliere N (2007) Windows anti-debug reference, Sept 2007. http://www.symantec.com/connect/articles/windows-anti-debug-reference
144. Gagnon MN, Taylor S, Ghosh AK (2007) Software protection through anti-debugging. Secur Priv Mag IEEE 5:82–84
145. Rutkowska J (2006) Subverting Vista Kernel for Fun and Profit. Kuala Lumpur
146. Desclaux F (2005) Rr0d: the rasta ring 0 debugger. http://rr0d.droids-corp.org/
147. Linn C, Debray S (2003) Obfuscation of executable code to improve resistance to static disassembly. Proceedings of the 10th ACM conference on computer and communications security, Washington, DC, USA, ACM, 2003, pp 290–299
148. Jacob M, Jakubowski MH, Venkatesan R (2007) Towards integral binary execution: implementing oblivious hashing using overlapped instruction encodings. Proceedings of the 9th workshop on multimedia security, Dallas, Texas, USA, ACM, 2007, pp 129–140
149. Collberg C, Thomborson C, Low D (1998) Manufacturing cheap, resilient, and stealthy opaque constructs. Proceedings of the 25th ACM SIGPLAN-SIGACT symposium on principles of programming languages, San Diego, CA, USA, ACM, 1998, pp 184–196
150. Ge J, Chaudhuri S, Tyagi A (2005) Control flow based obfuscation. Proceedings of the 5th ACM workshop on digital rights management, Alexandria, VA, USA, ACM, 2005, pp 83–92. http://portal.acm.org/citation.cfm?id=1102546.1102561
151. Cappaert J, Preneel B (2010) A general model for hiding control flow. USA, Chicago
152. Halderman A, Schoen S, Heninger N, Clarkson W, Paul W, Calandrino J, Feldman A, Appelbaum J, Felten E (2008) Lest we remember: cold boot attacks on encryption keys. Center for Information Security, Princeton University, Febr 2008. http://citp.princeton.edu/memory/
153. Shamir A, van Someren N (1999) Playing 'hide and seek' with stored keys. Proceedings of the financial cryptography
154. Cappaert J, Kisserli N, Schellekens D, Preneel B (2006) Self-encrypting code to protect against analysis and tampering, Novemb 2006. https://www.cosic.esat.kuleuven.be/publications/article-811.pdf
155. Boneh , DeMillo R, Lipton R (1997) On the importance of checking cryptographic protocols for faults. Advances in cryptology—eurocrypt, pp 37–51. http://dx.doi.org/10.1007/3-540-69053-0_4
156. Software Protection and Software Piracy Products—Arxan Technologies. http://www.arxan.com/software-protection/software-piracy-protection.php
157. MetaFortress Keeping Software Safe—Metaforic. http://www.metaforic.com/content/default.asp.

158. Anckaert B, Sutter BD, Bosschere KD (2004) Software piracy prevention through diversity. Proceedings of the 4th ACM workshop on digital rights management, Washington, DC, USA, ACM, 2004, pp 63–71. http://portal.acm.org/citation.cfm?id=1029146.1029157
159. Cohen FB (1993) Operating system protection through program evolution. Comput Secur 12: 565–584. http://www.sciencedirect.com/science/article/B6V8G-45K0N6R-HT/2/e4c9998c81912608a2d2b31d31ee361a
160. Schneier B (2000) Trusted client software. Crypto-gram, May 2000. http://www.schneier.com/crypto-gram-0005.html#6
161. Stefik M (1996) Letting loose the light: Igniting commerce in electronic publication, Xerox Parc, 1996. http://www2.parc.com/istl/projects/uir/publications/items/UIR-1996-10-Stefik-InternetCommerce-IgnitingDreams.pdf
162. Stefik M, Shifting the possible: how trusted systems and digital property rights challenge us to rethink digital publishing. http://www.law.berkeley.edu/journals/btlj/articles/vol12/Stefik/html/text.html
163. Stefik M (1998) Digital property rights language. Manual and Tutorial—XML Edition, Novemb 1998. http://xml.coverpages.org/DPRLmanual-XML2.html
164. eXtensible rights Markup Language (XrML) 2.0 specifications, 2001. http://www.xrml/org/
165. Contentguard, XrML 2.0 Technical Overview Version 1.0, March 2002
166. Bormans J, Hill K (2002) MPEG-21 overview V.5, ISO. http://mpeg.chiariglione.org/standards/mpeg-21/mpeg-21.htm
167. Iannella R (2002) Open digital rights language (ODRL) version 1.1, Sept 2002. http://www.w3.org/TR/odrl/
168. The Open Digital Rights Language Initiative. http://odrl.net/2.0/
169. ODRL: A Rights Expression Language for Digital Asset Management and E-Commerce. http://odrl.net/docs/ODRL-brochure.pdf
170. Chong CN, Corin R, Doumen J, Etalle S, Hartel P, Law YW, Tokmakoff A (2006) LicenseScript: a logical language for digital rights management. Annales des télécommunications, vol 61, 2006, pp 284–331. http://www.ub.utwente.nl/webdocs/ctit/1/00000122.pdf
171. Chong CN, Etalle S, Hartel PH, Law YW (2003) Approximating fair use in LicenseScript. Digital libraries: technology and management of indigenous knowledge for global access. Springer, Berlin, pp 432–443
172. Cheng S, Rambhia A (2003) DRM and standardization—can DRM be standardized? Digital rights management: technological, economic, legal aspects. Springer, Berlin
173. Content Protection & Copy Management Specification; Part 13: DVB-CPCM Compliance Framework (2008)
174. Marks D (2004) The role of Hook IP in content protection technology licensing, DVB-CPT
175. Microsoft (2004) Compliance and Robustness Rules for Windows Media DRM Implementations
176. Bramhill I, Sims M (1997) Challenges for copyright in a digital age. BT Technol J 15:63–73. http://www.ingentaconnect.com/content/klu/bttj/1997/00000015/00000002/00176332
177. Anti-counterfeiting Trade Agreement—Deliberative Draft, January 2010. http://www.laquadrature.net/wiki/ACTA_20100118_version_consolidated_text
178. DVD-copying software firm blocked again, out-law.com, May 2004. http://www.out-law.com/page-4571
179. Booch G, Rumbaugh J, Jacobson I (1998) The unified modeling language user guide. Addison-Wesley Professional, Reading
180. Iannella R (2001) Digital rights management (DRM) architectures. D-Lib Mag 7, June 2001. http://www.dlib.org/dlib/june01/iannella/06iannella.html
181. MultimediaWiki. http://wiki.multimedia.cx/index.php?title=Main_Page
182. Mauthe AU, Thomas DP (2004) Professional content management system: handling digital media assets. Wiley, Milton

183. Guth S (2006) Interoperability of DRM systems: Via the exchange of XML-based rights expressions. Peter Lang, Bern
184. Civolution, Nexguard Content Protection. http://www.civolution.com/applications/nexguard/content-protection/
185. Courtay O, Karroumi M (2007) AACS under fire. Secur Newsl 2, January 2007. http://eric-diehl.com/newsletterEn.html
186. Mulligan DK, Han J, Burstein AJ (2003) How DRM-based content delivery systems disrupt expectations of personal use. Proceedings of the 3rd ACM workshop on digital rights management, Washington, DC, USA, ACM, 2003, pp 77–89
187. Robinson S (1999) CD software Is said to monitor users' listening habits. New York Times, Novemb 1999. http://www.nytimes.com/library/tech/99/11/biztech/articles/01real.html
188. EMC Documentum Information Rights Management: overview of technical architecture, Dec 2006. http://www.emc.com/collateral/software/white-papers/h3395_irm_tech_overview_wp.pdf
189. Cuppens F, Miège A (2003) Modelling contexts in the Or-BAC Model. Proceedings of the 19th annual computer security applications Conference, 2003, p 416
190. Oracle, Information Rights Management—Managing Information Everywhere It Is Stored and Used, April 2007
191. Rosenblatt W, Mooney S, Trippe W (2001) Digital rights management: business and technology. Wiley, New York
192. Anderson R (2008) Security engineering: a guide to building dependable distributed systems. Wiley, New York
193. Jamkhedkar PA, Heileman GL (2004) DRM as a layered system. Proceedings of the 4th ACM workshop on digital rights management, Washington, DC, USA, ACM, 2004, pp 11–21
194. Heileman GL, Jamkhedkar PA (2005) DRM interoperability analysis from the perspective of a layered framework. Proceedings of the 5th ACM workshop on digital rights management, Alexandria, VA, USA, ACM, 2005, pp 17–26. http://portal.acm.org/citation.cfm?id=1102546.1102551
195. Diehl E (2008) A four-layer model for security of digital rights management. Proceedings of the 8th ACM workshop on digital rights management, Alexandria, Virginia, USA, ACM, 2008, pp 19–28
196. Decentralized DRM: Next Generation of DRMs, July 2004
197. Diehl E (2008) Trust & digital rights management, Rennes, France. http://eric-diehl.com/publications/CESAR08.pps
198. Anderson R, Kuhn M (1996) Tamper resistance: a cautionary note. Proceedings of the 2nd USENIX workshop on electronic commerce, Oakland, CA, USA, USENIX Association, 1996. http://portal.acm.org/citation.cfm?id=1267168&dl=
199. McCormac J (1996) European scrambling system: circuits, tactics and techniques: the black book, Baylin
200. Pohlmann KC (1992) The compact disc handbook. A-R Editions, Madison
201. Semeria C, Maufer T (1996) Introduction to IP multicast routing. http://www4.ncsu.edu/~rhee/export/papers/multi1.pdf
202. Andreaux J-P, Durand A, Furon T, Diehl E (2004) Copy protection system for digital home networks. Signal Process Mag IEEE 21:100–108. http://ieeexplore.ieee.org/xpls/abs_all.jsp?arnumber=1276118
203. FCC 04-193 Digital Output Protection Technology and Recording Method Certifications, August 2004. http://www.ics.uci.edu/~sjordan/courses/fall_design/FCC04-193.pdf
204. American Library Association v. Federal Communications Commission (2005) http://openjurist.org/406/f3d/689/american-library-association-v-federal-communications-commission
205. Platt C (2004) Satellite pirates. Wired

206. Lenoir V (1991) Eurocrypt, a successful conditional access system. IEEE international conference on consumer electronics, Rosemont, IL, pp 206–207. http://ieeexplore.ieee.org/Xplore/login.jsp?url=http%3A%2F%2Fieeexplore.ieee.org%2Fiel4%2F5591%2F14973%2F00733160.pdf%3Farnumber%3D733160&authDecision=-203
207. Leduc M (1990) Système de télévision à péage à contrôle d'accès pleinement détachable, un exemple d'implémentation: Videocrypt. Proceedings of ACSA, Rennes
208. Bower A, Clarke C, Robinson A (1995) LINE SHUFFLING: development of a scrambling system for terrestrial UHF television broadcasts. The British Broadcasting Corporation. http://downloads.bbc.co.uk/rd/pubs/reports/1995-11.pdf
209. Macq BM, Quisquater J-J (1995) Cryptology for digital TV broadcasting. Proc IEEE 83:944–957
210. ETR289 Digital Video Broadcasting (DVB) Support for use of scrambling and Conditional Access (CA) within digital broadcasting systems (1996) http://www.pdfdownload.org/pdf2html/pdf2html.php?url=http%3A%2F%2Fbroadcasting.ru%2Fpdf-standard-specifications%2Fconditional-access%2Fdvb-csa%2Fetr289-e1.pdf&images=yes
211. Le Buhan C (2006) Content Bulk Encryption for DRM Engineering, January 2006. http://www.cptwg.org/Assets/Presentations%202006/ScramblingForDRM.pps
212. Furht B, Muharemagic E, Socek D (2005) Multimedia encryption and watermarking. Springer, Heidelberg
213. Massoudi A, Lefebvre F, De Vleeschouwer C, Macq B, Quisquater J-J (2008) Overview on selective encryption of image and video: xhallenges and perspectives. http://dx.doi.org/10.1155/2008/179290
214. Support for use of the DVB Scrambling Algorithm version 3 within digital broadcasting systems, July 2008. http://www.dvb.org/(RoxenUserID=1eeee9e02e46b5b1b49ba0c982fe4ead)/technology/standards/a125_CSA3_dTR101289.v1.2.1.pdf
215. DVB—Digital Video Broadcasting—Home. http://www.dvb.org/
216. Press release: DVB agrees final elements of conditional access package, March 1995. http://www.dvb.org/documents/press-releases/pr008_Final%20Elements%20of%20Conditional%20Access%20Package%20Agreed.950307.pdf
217. DVB A215: Support for use of the DVB scrambling algorithm version 3 within digital broadcasting systems, July 2008
218. Weinmann R-P, Wirt K, Analysis of the DVB common scrambling algorithm. http://citeseer.ist.psu.edu/weinmann04analysis.html
219. Li W, Gu D (2007) Security analysis of DVB common scrambling algorithm. First international symposium on data, privacy, and e-commerce, ISDPE 2007, pp 271–273
220. DirecTV DSS Glossary of Terms. http://www.websitesrcg.com/dss/Glossary.htm
221. CENELEC, EN50221: common interface specification for conditional access and other digital video broadcasting decoder applications, Febr 1997. http://www.dvb.org/technology/standards/En50221.V1.pdf
222. Directive 2002/22/EC: on universal service and users' rights relating to electronic communications networks and services (Universal Service Directive), March 2002. http://eur-lex.europa.eu/LexUriServ/LexUriServ.do?uri=OJ:L:2002:108:0051:0077:EN:PDF
223. CI+ Specification Content Security Extensions to the Common Interface, May 2008
224. ANSI/SCTE 28 2007 Host-Pod Interface. http://www.scte.org/documents/pdf/Standards/ANSISCTE282007.pdf
225. Hearty PJ (2006) Carriage of digital video and other services by cable in North America. Proc IEEE 94:148–157
226. Brown D (1989) Dynamic feedback arrangement scrambling technique keystream generator. US Patent 4860353, 22 August 1989. http://www.google.com/patents?id=7II6AAAAEBAJ&printsec=abstract&zoom=4&source=gbs_overview_r&cad=0
227. OpenCable specifications. http://www.cablelabs.com/opencable/specifications/
228. Abennett, "Microsoft's digital rights management hacked," ITWorld, Oct 2001. http://www.itworld.com/IDG011024microsofthack

229. Leyden J (2005) Trojans exploit Windows DRM loophole. The Register, January 2005. http://www.theregister.co.uk/2005/01/13/drm_trojan/
230. Screamer B, Beale Screamer on Sci.crypt, cryptnome. http://cryptome.org/beale-sci-crypt.htm
231. Microsoft PlayReady Content Access Technology: White Paper, July 2008. http://download.microsoft.com/download/b/8/3/b8316f44-e7a9-48ff-b03a-44fb92a73904/Microsoft%20PlayReady%20Content%20Access%20Technology-Whitepaper.docx
232. The Official Microsoft Silverlight Site. http://www.silverlight.net/
233. Moonlight. http://www.mono-project.com/Moonlight
234. Jobs S (2007) Apple—thoughts on music, Febr 2007. http://www.apple.com/hotnews/thoughtsonmusic/
235. Doctorow C (2006) Opinion: Apple's copy protection isn't just bad for consumers, it's bad for business. Information Week, July 2006. http://www.informationweek.com/news/global-cio/showArticle.jhtml?articleID=191000408
236. Gassend B, Clarke D, van Dijk M, Devadas S (2002) Silicon physical random functions. Proceedings of the 9th ACM conference on computer and communications security, Washington, DC, USA, ACM, 2002, pp 148–160. http://portal.acm.org/citation.cfm?id=586110.586132
237. Halderman A (2005) Hidden feature in Sony DRM uses open source code to add Apple DRM. Freedom to Tinker, Dec 2005. http://www.freedom-to-tinker.com/blog/jhalderm/hidden-feature-sony-drm-uses-open-source-code-add-apple-drm
238. Adobe flash access: usage rules, May 2010. http://www.adobe.com/go/learn_flashaccess_usagerules_2
239. OMA home. http://www.openmobilealliance.org/
240. Open Mobile Alliance Ltd., DRM Architecture Version 2.0.1, Febr 2008
241. Adams C, Malpani A, Myers M, Galperin S, Ankney R (1999) X.509 internet public key infrastructure online certificate status protocol—OCSP, June 1999. http://tools.ietf.org/html/rfc2560
242. CMLA technical specifications (Dec 2005). http://www.cm-la.com/licensing/specifications.aspx
243. Marlin Overview Architecture (2007) https://www.marlin-community.com/public/MarlinArchitectureOverview.pdf
244. Boccon-Gibod G, Boeuf J, Lacy J (2009) Octopus: an application independent DRM toolkit. Proceedings of the 6th IEEE conference on consumer communications and networking, Las Vegas, NV, USA. IEEE Press, 2009, pp 1148–1154. http://portal.acm.org/citation.cfm?id=1700527.1700809
245. NEMO_Specifications.v.1.1.1, 2010. http://www.intertrust.com/system/files/NEMO_Specifications.v.1.1.1.pdf
246. Kamperman F, Szostek L, Baks W (2007) Marlin common domain: authorized domains in Marlin technology. 4th IEEE consumer communications and networking conference, 2007, pp 935–939
247. Intertrust, Seacert Trust Services. http://www.intertrust.com/solutions/seacert
248. Rassool R (2007) Widevine's Mensor. Broadcast Engineering, August 2007. http://broadcastengineering.com/products/widevines-mensor-watermarking-system/
249. Widevine to be Acquired by Google, Dec 2010. http://www.widevine.com/pr/206.html
250. Lawler R (2010) 5 reasons Google bought Widevine, Gigaom, Dec 2010. http://gigaom.com/video/google-widevine-acquisition/
251. Audio recording—compact disc digital audio system—Ed 2.0 (1999)
252. Sony DADC—digital solutions and production of CD, DVD, UMD, DualDisc and Blu-ray Disc. http://www.sonydadc.com/opencms/opencms/sites/sony/index.html
253. CD crack: magic marker indeed, Wired, May 2002. http://www.wired.com/science/discoveries/news/2002/05/52665
254. Information regarding WCP and MediaMax content protection. http://cp.sonybmg.com/xcp/

255. Sony BMG Litigation Info | Electronic Frontier Foundation. http://www.eff.org/cases/sony-bmg-litigation-info
256. Sparks S, Butler J (2005) 'Shadow Walker': raising the bar for rootkit detection, Las Vegas, NV, USA, 2005. http://blackhat.com/presentations/bh-usa-05/bh-us-05-sparks.pdf
257. Russinovich M (2005) Sony, Rootkits and digital rights management gone too far. Mark's Blog, Oct 2005. http://blogs.technet.com/markrussinovich/archive/2005/10/31/sony-rootkits-and-digital-rights-management-gone-too-far.aspx
258. Halderman J (2003) Evaluating new copy-prevention techniques for audio CDs. Digital rights management, 2003, pp 101–117. http://www.springerlink.com/content/5n3e9ajcw0pfa9qd
259. Sony, ARccOS—DVD-Video digital rights management, Sony DADC. http://www.sony dadc.com/opencms/opencms/sites/eu/en/Products/Copy_Control_Solutions/ARccOS.html
260. Doyle W, Copying copy protected optical discs. US Patent 20060245320A1. http://www.google.com/patents?id=FMKZAAAAEBAJ&printsec=abstract&zoom=4&source=gbs_overview_r&cad=0#v=onepage&q=&f=false
261. Halderman J (2003) Analysis of the MediaMax CD3 copy-prevention system. Princeton University, 2003. http://www.cs.princeton.edu/research/techreps/TR-679-03
262. Perrin C (2010) No autorun can help protect Microsoft Windows from malware. IT Security, Novemb 2010. http://blogs.techrepublic.com.com/security/?p=4670&tag=nl.e036
263. Borland J (2003) Shift key breaks CD copy locks. CNET news.com, Oct 2003. http://news.cnet.com/2100-1025_3-5087875.html
264. Fox B (2001) Fans get free replacement of copy-protected CD. New Scientist, Novemb 2001. http://www.newscientist.com/article/dn1586
265. Copy protection technical working group home page. http://www.cptwg.org/
266. Superior Court of the State of California, Complaint for injunctive relief for misappropriation of trade secrets, 1999. http://rgendeinedomain.de/decss/legal/doc00003.doc
267. Gallery of CSS Descramblers. http://www.cs.cmu.edu/ ~ dst/DeCSS/Gallery/
268. Smith A (2004) Cryptography regulations, Oct 2004. http://www.scribd.com/doc/754/Cryptography-Regulations
269. Stevenson F (1999) Cryptanalysis of contents scrambling system, Novemb 1999. http://www.cs.cmu.edu/ ~ dst/DeCSS/FrankStevenson/analysis.html
270. Parker D (1999) Cease and DeCSS: DVD's encryption code cracked—technology information. Emedia Professional, Novemb 1999. http://findarticles.com/p/articles/mi_m0FXG/is_12_12/ai_63973528/?tag=content;col1
271. Anderson R (2001) Why information security is hard—an economic perspective. http://citeseerx.ist.psu.edu/viewdoc/summary?doi=?doi=10.1.1.28.4054
272. Sony develops copyright protection solutions for digital music content, Febr 1999. http://www.sony.com/SCA/press/990225.shtml
273. Memory Stick Copyright ProteTechnology—MagicGate. http://www.sony.net/Products/SC-HP/cx_news/vol20/pdf/tw.pdf
274. Fiat A, Naor M (1994) Broadcast encryption, advances in cryptology—Crypto 1993. Lecturer notes in computer science, vol 773. Springer, Berlin, pp 480–491
275. Zeng W, Yu HH, Lin C-Y (2006) Multimedia security technologies for digital rights management. Academic Press, Amsterdam
276. 4C entity—welcome. http://www.4centity.com/
277. Naor D, Naor M, Lotspiech J (2001) Revocation and tracing routines for stateless receivers. Lecture notes in computer science, advances in cryptology, vol. 2139. Springer, Berlin
278. Prigent N, Karroumi M, Chauvin M, Eluard M (2007) AACS, nouveau standard de protection des contenus pré-enregistrés haute-définition. http://www.neekoonthe.net/docs/CESAR-AACS.pdf
279. ATARI Vampire, WinDVD 8 Device Key Found! Doom9, Febr 2007. http://forum.doom9.org/showthread.php?t=122664

280. Jin H, Lotspiech J, Meggido N (2006) Efficient traitor tracing, Oct 2006
281. Kocher P, Jaffe J, Jun B, Laren C, Lawson N (2003) Self-protecting digital content: a technical report from the CRI content security research initiative. Whitepaper
282. Zou D, Bloom JA (2009) H.264/AVC substitution watermarking: a CAVLC example. Proceedings of the SPIE, San Jose, CA, USA, 2009. http://link.aip.org/link/PSISDG/v7254/i1/p72540Z/s1&Agg=doi
283. Cinavia Consumer Information Center. http://www.cinavia.com/languages/english/index.html
284. AACS for Blu-ray Disc Pre-recorded Book, v0.951, January 2010. http://www.aacsla.com/specifications/AACS_Spec_BD_Prerecorded_Final_0.951.pdf
285. Hitachi, Intel, Sony, Toshiba, Matsushita Electric Industrial, digital transmission content protection specification, vol 1 (Informational Version), Oct 2007. http://www.dtcp.com/data/info%2020071001%20DTCP%20V1%201p51.pdf
286. Kelsey J, Schneier B, Wagner D (1999) Mod n cryptanalysis, with applications against RC5P and M6. Lecture notes in computer science, pp 139–155
287. High-bandwidth digital content protection system Review 1.4, July 2009. http://www.digital-cp.com/files/static_page_files/5C3DC13B-9F6B-D82E-D77D8ACA08A448BF/HDCP%20Specification%20Rev1_4.pdf
288. Sichuan Changhong Electric Ltd., UCPS China's content protection specification for Digital Interface
289. Crosby S, Goldberg I, Johnson R, Song D, Wagner D (2002) A cryptanalysis of the high-bandwidth digital content protection system. Security and privacy in digital rights management. Springer, 2002. http://www.springerlink.com/content/x82b2g9tvuj8vv64/
290. Irwin K (2001) Four simple cryptographic attacks on HDCP, August 2001. http://www.angelfire.com/realm/keithirwin/HDCPAttacks.html
291. HDCP MASTER KEY! Pastebin.com. http://pastebin.ca/2018687
292. Content Protection & Copy Management Specification, July 2008. http://www.dvb.org/technology/standards/#content
293. Digital Living Network Alliance. http://www.dlna.org/home
294. EMI, Copy control—Faire une copie analogique. http://copycontrol.emi-artistes.com/copie.html
295. doubleTwist. http://www.doubletwist.com/dt/Home/Index.dt.
296. Diehl E, Furon T (2003) Copyright watermark: closing the analog hole. Proceedings of the IEEE international conference on consumer electronics, 2003, pp 52–53
297. Furon T (2007) A constructive and unifying framework for zero-bit watermarking. IEEE Trans Inf Forensics Secur 2:149–163
298. Talstra J, Justin M, Maes J-B, Lokhoff G, Import control of content. US Patent US2006 0075242
299. Wu M, Craver S, Felten EW, Liu B (2001) Analysis of attacks on SDMI audio watermarks. IEEE international conference on acoustics, speech, and signal processing (ICASSP '01), Salt Lake City, USA, IEEE, 2001, pp 1369–1372
300. BOWS-2 Web page. http://bows2.gipsa-lab.inpg.fr/
301. Craver S, Atakli I, Yu J (2007) How we broke the BOWS watermark, p 46. http://adsabs.harvard.edu/abs/2007SPIE.6505E..46C
302. Bloom JA (2003) Security and rights management in digital cinema. Proceedings of the IEEE international conference on acoustics, speech, and signal processing (ICASSP) 2003, vol 4, pp IV-712-15
303. NexGuard—Digital Cinema, Novemb 2010. http://www.civolution.com/fileadmin/bestanden/NexGuard%20-%20Digital%20Cinema/NexGuard_-_Digital_Cinema_-_nov_2010.pdf
304. Baudry S, Correa P, Doyen D, Macq BMM, de Manerffe O, Mas J-M, Nguyen P, Nivart JF, Wolf I (2007) European digital cinema security white book. Presses Universitaires de Louvain, Louvain-la-Neuve

305. Viewers upset over digital TV taping restrictions. The Japan Times Online, May 2004. http://search.japantimes.co.jp/member/member.html?nb20040525a2.htm
306. Ghosemajumder S (2002) Advanced peer-based technology business models. Sloan School of Management. http://hdl.handle.net/1721.1/8438
307. Kingsley-Hughes A (2007) Microsoft rebrands PlaysForSure, gives consumers a reason to buy iPods. Hardware 2.0, Dec 2007. http://blogs.zdnet.com/hardware/?p=1034&tag=nl.e622
308. Healey J (2008) Yahoo pulls an MSN Music (only faster). BitPlayer, July 2008. http://opinion.latimes.com/bitplayer/2008/07/yahoo-pulls-and.html
309. Sandoval G (2008) Interview: Microsoft's Rob Bennett defends DRM decision. Tech news blog, April 2008. http://www.news.com/8301-10784_3-9926741-7.html
310. Center for Content Protection. Technology issues on the use of DRM, July 2008
311. Schmidt AU, Tafreschi O, Wolf R, Interoperability challenges for DRM systems. http://citeseer.ist.psu.edu/744361.html
312. Bomsel O (2007) Gratuit! Du déploiement de l'économie numérique. Gallimard, Paris
313. Reuters, Music downloaders hit by acronym cacophony. CNET news.com, July 2004. http://www.news.com/Music-downloaders-hit-by-acronym-cacophony/2100-1027_3-5257842.html
314. Safavi-Naini R, Sheppard NP, Uehara T (2004) Import/export in digital rights management. Proceedings of the 4th ACM workshop on digital rights management, 2004, pp 99–110
315. Diehl E (2004) MediaNet: aframework to unify different distribution channels, Oct 2004. http://www.ist-ipmedianet.org/Medianet_position_paper_final.pdf
316. Koenen RH, Lacy J, MacKay M, Mitchell S (2004) The long march to interoperable digital rights management. Proc IEEE 92:883–897
317. Larose G, DAM and interoperable DRM: maintaining agility in the world of evolving content ecosystems. http://www.ingentaconnect.com/content/pal/dam/2006/00000002/00000001/art00004
318. DVB, ETSI TS 101 197: DVB SimulCrypt; Head-end architecture and synchronization, Febr 2002. http://www.etsi.org/deliver/etsi_ts/101100_101199/101197/01.02.01_60/ts_101197v010201p.pdf
319. Wirt K (2005) Fault attack on the DVB common scrambling algorithm. Computational science and Its applications, ICCSA, Lecture notes in computer science. Springer, Berlin, pp 511–577
320. DReaM-CAS Intellectual Property Review, Dec 2006. http://www.openmediacommons.org/collateral/DReaM-CAS_IPR_White_Paper_v1.0.pdf
321. Lee J, An CP, Laurent ML, Siedle D (2008) OpenCA: conditional access system in MediaFLOTM. IEEE international symposium on broadband multimedia systems and broadcasting, Las Vegas, USA, pp 1–6
322. Piracy: the good, the bad and the ugly. BroadcastPro, Novemb 2010. http://broadcastprome.com/content.aspx?sID=21&fID=128
323. SVP Open Content Protection Technical Overview, January 2005. http://www.svpalliance.org/docs/e2e_technical_introduction.pdf
324. Lee P, Necula G (1997) Research on proof-carrying code for mobile-code security. DARPA workshop on foundations for secure mobile code, pp 26–28
325. ISO/IEC MPEG4 IPMP overview & applications document, Dec 1998
326. Commerce Secretary Announces New Standard for Global Information Security, Dec 2001. http://www.nist.gov/public_affairs/releases/g01-111.htm
327. Diehl E, Robert A (2010) An introduction to interoperable rights locker. Proceedings of the annual ACM workshop on digital rights management (DRM), Chicago, USA, ACM, 2010, pp 51–54
328. Martin J (2010) How ultraviolet's DRM could impact the wireless industry. FierceMobileContent, Oct 2010. http://www.fiercemobilecontent.com/node/13395/print
329. Welcome to Coral Consortium. http://www.coral-interop.org/

330. Kalker T, Carey K, Lacy J, Rosner M (2007) The Coral DRM interoperability framework, Las Vegas, NV, USA, 2007, pp 930–934. http://www.intertrust.com/download/ccnc07.pdf
331. Hewlett-Packard, Intertrust, Panasonic, Philips, Samsung, Sony, and Universal Music Group. The Coral DRM interoperability framework applied to the DNLA/CPS DRM interoperability solution, March 2008
332. Bocharov J, Burns Q, Folta F, Hughes K, Murching A, Olson L, Schnell P, Simmons J (2009) Portable encoding of audio-video objects, Sept 2009. http://www.iis.net/downloads/files/media/smoothspecs/ProtectedInteroperableFileFormat(PIFF).pd
333. Microsoft Community Promise. Microsoft Interoperability, Sept 2007. http://www.microsoft.com/interop/cp/default.mspx
334. ISO—International Organization for Standardization. ISO/IEC 14496-12:2008 Information technology—coding of audio-visual objects—part 12: ISO base media file format. http://www.iso.org/iso/iso_catalogue/catalogue_ics/catalogue_detail_ics.htm?csnumber=51533
335. Netflix Taps Microsoft PlayReady as Its Primary DRM Technology for Netflix Ready Devices and Applications, May 2010. http://www.microsoft.com/presspass/press/2010/may10/05-25playreadynetflixpr.mspx
336. Diehl E (2008) Be our guest: Ton Kalker. Secur Newsl. http://eric-diehl.com/letter/Security%20Newsletter%20Summer%2008.pdf
337. Digital Entertainment Content Ecosystem (DECE) Announces Key Milestones, January 2010. http://www.uvvu.com/press/CES_2010_Press_Release.pdf
338. Galuten A (2010) DRM Interoperability Marlin, CORAL & DECE, April 2010. http://www.itu.int/dms_pub/itu-t/oth/15/08/T15080000040001PDFE.pdf
339. Smith E (2009) Disney touts a way to ditch the DVD. Wall Street J, Oct 2009. http://online.wsj.com/article/SB10001424052748703816204574485650026945222.html
340. Disney Rethinks Role of Keychest in Film Distribution. PaidContent, Novemb 2010. http://paidcontent.org/article/419-exclusive-sorry-keychest.-theres-a-new-disney-plan-for-film-distributio/
341. DMP Home Page. http://www.digital-media-project.org/
342. Chillout. http://chillout.dmpf.org/
343. Approved Document No 2—Technical Reference: Architecture—Version 3.2, Oct 2008. http://open.dmpf.org//dmp1202.zip
344. OpenSSL. http://www.openssl.org/
345. The Open Source Definition. Open Source Initiative. http://opensource.org/docs/osd
346. Home: AACS—Advanced Access Content System. http://www.aacsla.com/home
347. phundie (2008) Eavesdropping with LD_PRELOAD, 2600 the hacker quarterly, Spring 2008, pp 20–21
348. Liu Q, Safavi-Naini R, Sheppard NP (2003) Digital rights management for content distribution. Proceedings of the Australasian information security workshop on ACSW Frontiers, Adelaide, Australia. Australian Computer Society, 2003, pp 49–58. http://portal.acm.org/citation.cfm?id=827987.827994
349. Fernando G, Jacobs T, Swatinathan V (2005) Project DReaM: an architectural overview, Sept 2005
350. Welte J (2006) Yahoo, EMI, Norah Jones test MP3 sale. MP3.com, Dec 2006. http://www.mp3.com/news/stories/7506.html
351. Digital Watermarking Alliance. Digital watermarking alliance supports music industry efforts to sell MP3 music downloads and enhance consumer experiences, Dec 2006. http://www.digitalwatermarkingalliance.org/pr_121806.asp
352. Rosenblatt W (2008) Content identification technologies, April 2008. http://www.giantstepsmts.com/Content%20ID%20Whitepaper.pdf
353. Privacy principles for digital watermarking, May 2008. http://www.cdt.org/
354. Kalker T, Epema DHJ, Hartel PH, Lagendijk RL, Van Steen M (2004) Music2Share—copyright-compliant music sharing in P2P systems. Proc IEEE 92:961–970

355. Global File Registry White Paper, May 2006. http://www.globalfileregistry.com/assets/Global_File_Registry_White_Paper.pdf
356. Romp G (1997) Game theory: introduction and applications. Oxford University Press, Oxford
357. Heileman GL, Jamkhedkar PA, Khoury J, Hrncir CJ (2007) The DRM game. Proceedings of the ACM workshop on digital rights management, Alexandria, Virginia, USA, ACM, 2007, pp 54–62. http://portal.acm.org/citation.cfm?id=1314287
358. Hughes D, Coulson G, Walkerdine J (2005) Free riding on Gnutella revisited: the bell tolls? IEEE distributed systems Online, vol 6
359. RIAA Watch. http://sharenomore.blogspot.com/
360. Diehl E (2009) 80,000$ per song. The blog of content security, June 2009. http://eric-diehl.com/blog/?x=entry:entry090622-175234
361. Anderson N (2008) RIAA doubles settlement cost for students fighting subpoenas. Arstechnica, June 2008. http://arstechnica.com/news.ars/post/20080611-riaa-doubles-settlement-cost-for-students-fighting-subpoenas.html
362. Diehl E (2010) Ubisoft's DRM torpedoed!. The blog of content protection, March 2010. http://eric-diehl.com/blog/?x=entry:entry100309-205343
363. Digital Music Report 2009, New business models for a changing environment, January 2009. http://www.ifpi.org/content/library/DMR2009.pdf
364. Wallenstein A (2011) New iTunes features aim to Outdo DVD. PaidContent, January 2011. http://paidcontent.org/article/419-new-itunes-features-aim-to-outdo-dvd/
365. Thomson, SmartRight: technical white paper, January 2003. http://www.smartright.org/images/SMR/content/SmartRight_tech_whitepaper_jan28.pdf